An engineer's view
of human error

An engineer's view of human error

Third edition

Trevor Kletz

The theme of this book:
Try to change situations, not people

The information in this book is given in good
faith and belief in its accuracy, but does not
imply the acceptance of any legal liability or
responsibility whatsoever, by the Institution,
or by the author, for the consequences of its
use or misuse in any particular circumstances.

Published by
Institution of Chemical Engineers (IChemE),
Davis Building,
165–189 Railway Terrace,
Rugby, Warwickshire CV21 3HQ, UK
IChemE is a Registered Charity

© 2001 Trevor Kletz
First edition 1985
Second edition 1991

ISBN 0 85295 430 1

Printed in the United Kingdom by Bell & Bain Limited, Glasgow

Foreword to the third edition

In this book I set down my views on human error as a cause of accidents and illustrate them by describing a number of accidents that have occurred, mainly in the oil and chemical industries. Though the book is particularly addressed to those who work in the process industries, I hope that it will interest all who design, construct, operate and maintain plant of all sorts, that it may suggest to some readers new ways of looking at accident prevention and that it will reinforce the views of those readers who already try to change situations rather than people. It is intended for practising engineers, especially chemical engineers rather than experts in human factors and ergonomics, and I hope it will be of particular interest to students.

Although human factors are not specifically mentioned in the Institution of Chemical Engineers' publication *Accreditation of University Degree Courses — A Guide for University Departments* (November 1996), they are essential if the course is to 'develop an awareness of the necessity of safe design', which is included. Many of the accidents I describe can be used as the raw material for discussions on the causes of accidents and the action needed to prevent them happening again, using the methods described in the Institution's safety training packages, collections of notes and slides on accidents that have occurred. Some of the incidents are included in the package on *Human Error* and other packages.

Some readers may wonder why I have added to the existing literature on human error. When so much has been written, is there need for more?

I felt that much of the existing literature is too theoretical for the audience I have in mind, or devoted to particular aspects of the problem. I felt there was a need for a book which would suggest to engineers how they might approach the problem of human error and do so by describing accidents which at first sight seem to be the result of human error. My approach is therefore pragmatic rather than theoretical, and influenced by engineering methods. Thus I have:

- questioned the accepted wisdom;
- started with situations as they are rather than as we would like them to be;
- judged possible actions by their effectiveness;
- suggested actions based on the best current information, as in industry we often cannot postpone decisions until more information is available.

I do not claim any great originality for the ideas in this book. Many of them are to be found in other books such as *Man-Machine Engineering* by A. Chapanis (Tavistock, 1965) but while that book is primarily concerned with mechanical and control equipment, this one is concerned mainly with process equipment.

Not all readers may agree with the actions I propose. If you do not, may I suggest that you decide what action you think should be taken. Please do not ignore the accidents. They happened and will happen again, unless action is taken to prevent them happening.

Thanks are due to the many friends and colleagues, past and present, who suggested ideas for this book or commented on the draft — without their contributions I could not have produced this book — and to Mr S. Coulson who prepared the illustrations for Chapter 14. Thanks are also due to the many companies who allowed me to describe their mistakes and to the Science and Engineering Research Council for financial support for the earlier editions.

In the second edition I added three new chapters, on 'Accidents that could be prevented by better management', on 'Errors in computer-controlled plants' (as people, not computers make the errors) and on 'Some final thoughts' and made a number of additions to the existing chapters.

In this third edition I have added more examples of accidents caused by the various types of human error. I have extended the chapters on errors made by managers and designers and now include errors due to their ignorance of various options such as inherently safer design. I have also expanded the chapter on computer control and added an appendix on 'Some myths of human error', widespread beliefs that are not wholly true. John Doe has been joined in his adventures by Joe Soap but the reasons for their errors are very different.

I have often been irritated by authors who use phrases such as, 'as discussed in an earlier chapter', without saying which one. I have therefore included cross-references whenever a topic is discussed under more than one heading.

US readers should note that in the UK 'supervisor' is another name for a foreman while 'manager' is used to describe anyone above the level of foreman, including those who are known as supervisors or superintendents in most US companies.

To avoid the clumsy phrases 'he or she' and 'him or her' I have usually used 'he' and 'him'. Though there has been a welcome increase in the number of women employed in the process industries, the manager, designer or accident victim is still usually a man.

Trevor Kletz
May 2001

Glossary

Some words are used in this book in a specialized sense.

Error
A failure to carry out a task in the way intended by the person performing it, in the way expected by other people or in a way that achieves the desired objective.

This definition has been worded so as to include the various types of error listed below. It is consistent with the *Shorter Oxford English Dictionary* which defines error as 'something incorrectly done through ignorance or inadvertence' and as 'a transgression'. Some writers use error in a narrower sense to include only slips and lapses of attention, as defined below.

Mismatch
An error that occurs because the task is beyond the physical or mental ability of the person asked to perform it, perhaps beyond anyone's ability (see Chapter 4).

Mistake
An error that occurs as a result of ignorance of the correct task or the correct way to perform it. The intention is fulfilled but the intention is wrong (see Chapter 3).

Non-compliance or violation
An error that occurs because someone decides not to carry out a task or not to carry it out in the way instructed or expected. The motive can range from sabotage, through 'can't be bothered' to a belief that the instructions were incorrect. In such cases the non-compliance may prevent a mistake (as defined above) (see Chapter 5).

Slips and lapses of attention
An error that occurs as a result of forgetfulness, habit, fatigue or similar psycho-logical causes. Compared with mistakes (as defined above), the intention is correct but it is not fulfilled (see Chapter 2).

The opening words of Tolstoy's *Anna Karenina* are, 'All happy families resemble one another, each unhappy family is unhappy in its own way.' Similarly, there is a right way for every task but each error is erroneous in its own way.

Contents

Foreword to the third edition   iii

1   Introduction   1
1.1   Accept men as we find them   1
1.2   Meccano or dolls?   3
1.3   Types of human error   4
1.4   Two simple examples   6
1.5   Accident investigation   8
1.6   A story   9
1.7   Research on human error   10

2   Accidents caused by simple slips   11
2.1   Introduction   11
2.2   Forgetting to open or close a valve   13
2.3   Operating the wrong valve   21
2.4   Pressing the wrong button   25
2.5   Failures to notice   29
2.6   Wrong connections   31
2.7   Errors in calculations   32
2.8   Other medical errors   34
2.9   Railways   35
2.10 Other industries   41
2.11 Everyday life (and typing)   43
2.12 Fatigue   45

3   Accidents that could be prevented by better training or instructions 48
3.1   Introduction   48
3.2   Three Mile Island   50
3.3   Other accidents that could be prevented by relatively sophisticated training 53
3.4   Accidents that could be prevented by elementary training   62
3.5   Contradictory instructions   65
3.6   Knowledge of what we don't know   66

3.7	Some simple ways of improving instructions	67
3.8	Training or instructions?	72
3.9	Cases when training is not the best answer	74

4	**Accidents due to a lack of physical or mental ability**	78
4.1	People asked to do the physically difficult or impossible	78
4.2	People asked to do the mentally difficult or impossible	80
4.3	Individual traits and accident proneness	85
4.4	Mind-sets	88

5	**Accidents due to failures to follow instructions**	97
5.1	Accidents due to non-compliance by managers	99
5.2	Accidents due to non-compliance by operators	103
5.3	Actions to improve compliance	107
5.4	Alienation	109
5.5	Postscript	110

6.	**Accidents that could by prevented by better management**	112
6.1	An accident caused by insularity	114
6.2	An accident due to amateurism	115
6.3	The fire at King's Cross railway station	116
6.4	The Herald of Free Enterprise	117
6.5	The Clapham Junction railway accident	118
6.6	Piper Alpha	120
6.7	What more can senior managers do?	120
6.8	The measurement of safety	125
6.9	Conclusions	127

7	**The probability of human error**	129
7.1	Why do we need to know human error rates?	130
7.2	Human error rates – a simple example	132
7.3	A more complex example	132
7.4	Other estimates of human error rates	136
7.5	Two more simple examples	141
7.6	Button pressing	144
7.7	Non-process operations	146
7.8	Train driver errors	147
7.9	Some pitfalls in using data on human reliability	147
7.10	Data on equipment may be data on people	149

| 7.11 | Who makes the errors? | 151 |
| 7.12 | Conclusions | 151 |

8	**Some accidents that could be prevented by better design**	154
8.1	Isolation of protective equipment	155
8.2	Better information display	156
8.3	Pipe failures	156
8.4	Vessel failures	159
8.5	The Sellafield leak	159
8.6	Other design errors	161
8.7	Conceptual shortcomings	162
8.8	Problems of design contractors	165
8.9	Domestic accidents	166

9	**Some accidents that could be prevented by better construction**	168
9.1	Pipe failures	168
9.2	Miscellaneous incidents	170
9.3	Prevention of construction errors	173

10	**Some accidents that could be prevented by better maintenance**	175
10.1	Incidents which occurred because people did not understand how equipment worked	175
10.2	Incidents which occurred because of poor maintenance practice	178
10.3	Incidents due to gross ignorance or incompetence	180
10.4	Incidents which occurred because people took short cuts	181
10.5	Incidents which could be prevented by more frequent or better maintenance	183
10.6	Can we avoid the need for so much maintenance?	185

11	**Some accidents that could be prevented by better methods of operation**	187
11.1	Permits-to-work	187
11.2	Tanker incidents	189
11.3	Some incidents that could be prevented by better instructions	192
11.4	Some incidents involving hoses	193
11.5	Communication failures	194
11.6	Examples from the railways	199
11.7	Simple causes in high tech industries	201

12	**Errors in computer-controlled plants**	203
12.1	Hardware failures	204
12.2	Software errors	205
12.3	Specification errors	207
12.4	Misjudging responses to a computer	210

12.5   Entering the wrong data   213
12.6   Failures to tell operators of changes in data or programs   214
12.7   Unauthorized interference with hardware or software   214
12.8   The hazards of old software   217
12.9   Other applications of computers   217
12.10 Conclusions   218

13   Personal and managerial responsibility   221
13.1   Personal responsibility   221
13.2   Legal views   223
13.3   Blame in accident investigations   227
13.4   Managerial wickedness   228
13.5   Managerial competence   229
13.6   Possible and necessary   231

14   The adventures of Joe Soap and John Doe   234

15   Some final thoughts   254

Postscript   257

Appendix 1 – Influences on morale   258

Appendix 2 – Some myths of human error   261

Appendix 3 – Some thoughts on sonata form   269

Further reading   271

Index   273

'... unlike all other organisms Homo sapiens *adapts not through modification of its gene pool to accommodate the environment but by manipulating the environment to accommodate the gene pool ...'*
M.T. Smith and R. Layton, 1989, *The Sciences*, 29(1): 10

'... there is a belief amongst many engineers and managers that human error is both inevitable and unpredictable. However, human error is inevitable only if people are placed in situations that emphasise human weaknesses and that do not support human strengths.'
Martin Anderson, *IChemE Safety and Loss Prevention Subject Group Newsletter*, Spring 1999

'... decreasing the amount of human responsibility in the operation of the plant increases the amount of human responsibility in the design of the plant.'
The Use of Computers in Safety-critical Applications (HSE Books, UK, 1998), page 13

'If the honey that the bees gather out of so manye floure of herbes ... that are growing in other mennis medowes ... may justly be called the bees' honeye ... so maye I call it that I have ... gathered of manye good autores ... my booke.'
William Turner (1510–1668). Quoted by Frank Lees in *Loss Prevention in the Process Industries*, 2nd edition (Butterworth-Heinemann, Oxford, UK, 1996) and reproduced here as a tribute to one who did so much to further the subject.

Introduction

1

'Man is a creature made at the end of the week ...
when God was tired.'
Mark Twain

1.1 Accept men as we find them

The theme of this book is that it is difficult for engineers to change human
nature and therefore, instead of trying to persuade people not to make errors, we
should accept people as we find them and try to remove opportunities for error
by changing the work situation — that is, the plant or equipment design or the
method of working. Alternatively, we can mitigate the consequences of error or
provide opportunities for recovery. (When it is possible for them to do so,
people are better at correcting errors than at not making them.) I hope the book
will remind engineers of some of the quirks of human nature so that they can
better allow for them in design.

The method used is to describe accidents which at first sight were due to
human error and then discuss the most effective ways of preventing them
happening again. The accidents occurred mainly, though not entirely, in the
oil and chemical industries, but nevertheless should interest all engineers, not
just chemical engineers, and indeed all those who work in design or production. Apart from their intrinsic interest, the accident reports will, I hope, grab
the reader's attention and encourage him or her to read on. They are also
more important than the advice. I did not collect incident reports to illustrate
or support my views on prevention. I developed my views as the result of
investigating accidents and reading accident reports. You may not agree with
my recommendations but you should not ignore the reports.

Browsing through old ICI files I came across a report dating from the late
1920s in which one of the company's first safety officers announced a new
discovery: after reading many accident reports he had realized that most accidents are due to human failing. The remedy was obvious. We must persuade
people to take more care.

Since then people have been exhorted to do just this, and this policy has
been supported by tables of accident statistics from many companies which
show that over 50%, sometimes as many as 90%, of industrial accidents are
due to human failing, meaning by that the failure of the injured man or a

1

fellow-worker. (Managers and designers, it seems, are not human or do not fail.) This is comforting for managers. It implies that there is little or nothing they can do to stop most accidents.

Many years ago, when I was a manager, not a safety adviser, I looked through a bunch of accident reports and realized that most of the accidents could be prevented by better management — sometimes by better design or method of working, sometimes by better training or instructions, sometimes by better enforcement of the instructions.

Together these may be called changing the work situation. There was, of course, an element of human failing in the accidents. They would not have occurred if someone had not forgotten to close a valve, looked where he was going, not taken a short-cut. But what chance do we have, without management action of some sort, of persuading people not to do these things?

To say that accidents are due to human failing is not so much untrue as unhelpful, for three reasons:

(1) Every accident is due to human error: someone, usually a manager, has to decide what to do; someone, usually a designer, has to decide how to do it; someone, usually an operator, has to do it. All of them can make errors but the operator is at the end of the chain and often gets all the blame. We should consider the people who have opportunities to prevent accidents by changing objectives and methods as well as those who actually carry out operations (see Appendix 2, item 1, page 261).

(2) Saying an accident is due to human failing is about as helpful as saying that a fall is due to gravity. It is true but it does not lead to constructive action. Instead it merely tempts us to tell someone to be more careful. But no-one is deliberately careless; telling people to take more care will not prevent an accident happening again. We should look for changes in design or methods of working that can prevent the accident happening again.

(3) The phrase 'human error' lumps together different sorts of failure that require different quite actions to prevent them happening again (see Section 1.3, page 4).

If all accidents are due to human errors, how does this book differ from any other book on accidents? It describes accidents which at first sight seem to be due wholly or mainly to human error, which at one time would have been followed by exhortations to take more care or follow the rules, and emphasizes what can be done by changing designs or methods of working. The latter phrase includes training, instructions, audits and enforcement as well as the way a task is performed.

It is better to say that an accident can be prevented by better design, better instructions, etc, than to say it was caused by bad design, instructions, etc. Cause implies blame and we become defensive. We do not like to admit that we did something badly, but we are willing to admit that we could do it better.

I do not say that it is impossible to change people's tendency to make errors. Those more qualified than engineers to do so — teachers, clergymen, social workers, psychologists — will no doubt continue to try and we wish them success. But the results achieved in the last few thousand years suggest that their results will be neither rapid nor spectacular and where experts achieve so little, engineers are likely to achieve less. Let us therefore accept that people are the one component of the systems we design that we cannot redesign or modify. We can design better pumps, compressors, distillation columns, etc, but we are left with Mark I man and woman.

We can, of course, change people's performance by better training and instructions, better supervision and, to some extent, by better motivation. What we cannot do is enable people to carry out tasks beyond their physical or mental abilities or prevent them making occasional slips or having lapses of attention. We can, however, reduce the opportunities for such slips and lapses of attention by changing designs or methods of working.

People are actually very reliable but there are many opportunities for error in the course of a day's work and when handling hazardous materials we can tolerate only very low error rates (and equipment failure rates), lower than it may be possible to achieve. We may be able to keep up a tip-top performance for an hour or two while playing a game or a piece of music but we cannot keep it up all day, every day. Whenever possible, therefore, we should design user-friendly plants which can tolerate human error (or equipment failure) without serious effects on safety, output or efficiency[3].

1.2 Meccano or dolls?

Let me emphasize that when I suggest changing the work situation, I am not simply saying change the hardware. Sometimes we have to change the software — the method of working, training, instructions, etc. Safety by design should always be the aim, but sometimes redesign is impossible, or too expensive, and we have to modify procedures. In over half the accidents that occur there is no reasonably practical way of preventing a repetition by a change in design and we have to change the software.

At present, most engineers are men and as boys most of us played with Meccano rather than dolls. We were interested in machines and the way they work, otherwise we would not be engineers. Most of us are very happy to devise hardware solutions. We are less happy when it comes to software solutions, to devising new training programmes or methods, writing instructions, persuading people to follow them, checking up to see that they are being followed and so on. However, these actions are just as important as the hardware ones, as we shall see, and require as much of our effort and attention.

One reason we are less happy with software solutions is that continual effort — what I have called grey hairs[1] — is needed to prevent them disappearing. If a hazard can be removed or controlled by modifying the hardware or installing extra hardware, we may have to fight for the money, but once we get it and the equipment is modified or installed it is unlikely to disappear.

In contrast, if a hazard is controlled by modifying a procedure or introducing extra training, we may have less difficulty getting approval, but the new procedure or training programme may vanish without trace in a few months once we lose interest. Procedures are subject to a form of corrosion more rapid and thorough than that which affects the steelwork. Procedures lapse, trainers leave and are not replaced. A continuous management effort — grey hairs — is needed to maintain our systems. No wonder we prefer safety by design whenever it is possible and economic; unfortunately, it is not always possible and economic.

Furthermore, when we do go for safety by design, the new equipment may have to be tested and maintained. It is easy to install new protective equipment — all you have to do is persuade someone to provide the money. You will get more grey hairs seeing that the equipment is tested and maintained and that people are trained to use it properly and do not try to disarm it.

1.3 Types of human error

Human errors occur for various reasons and different actions are needed to prevent or avoid the different sorts of error. Unfortunately much of the literature on human error groups together widely different phenomena which call for different action, as if a book on transport discussed jet travel and seaside donkey rides under the same headings (such as costs, maintenance and publicity). I find it useful to classify human errors as shown below. Most classification systems are designed primarily to help us find the information. I have used a system that helps us find the most effective way of preventing the accidents happening again.

- Errors due to a *slip* or momentary *lapse of attention* (discussed in Chapter 2).

The intention is correct but the wrong action or no action is taken. We should reduce opportunities for errors by changing the work situation.

- Errors due to poor training or instructions (discussed in Chapter 3). Someone does not know what to do or, worse, thinks he knows but does not. These are called *mistakes*. The intention is carried out but is wrong. We need to improve the training or instructions or simplify the job.
- Errors which occur because a task is beyond the physical or mental ability of the person asked to do it, perhaps beyond anyone's ability (discussed in Chapter 4). There is a *mismatch* between the ability of the person and the requirements of the task. We need to change the work situation.
- Errors due to a deliberate decision not to follow instructions or accepted practice (discussed in Chapter 5). These are often called *violations* but *non-compliance* is a better term, as people often believe that the rule is wrong or that circumstances justify an exception. We should ask why the rules were not followed. Did someone not understand the reasons for the instructions, were they difficult to follow, have supervisors turned a blind eye in the past? There is fine line between initiative and breaking the rules. Note that if the instructions were wrong, non-compliance may prevent a mistake (as defined above).

Chapter 6 discusses those errors made by managers, especially senior managers, because they do not realize that they could do more to prevent accidents. These errors are not a fifth category but are mainly due to ignorance of what is possible. Many managers do not realize that they need training and there is no-one who can tell them. Today it is widely recognized that all accidents are management failings: failures to reduce opportunities for error, failures to provide adequate training and instruction, failures to ensure that people follow the rules, and so on. Chapter 6, however, looks at management failures in a narrower sense.

The boundaries between these categories are not clear-cut and often more than one factor may be at work. Thus an error might be due to poor training compounded by limited ability (a mismatch) and the fact that the foreman turned a blind eye on previous occasions, thus winking at non-compliance.

Tasks are sometimes divided into skill-based, rule-based and knowledge-based. Skill-based actions are highly practised ones carried out automatically with little or no conscious monitoring. Knowledge-based actions are carried out under full conscious control, because the task is non-routine or unfamiliar; they take longer than skill-based ones as each step has to be considered. Rule-based actions have been learned but have not been carried out often enough to become automatic; the degree of conscious control is intermediate.

Note that the same action may be skill-based for one person, rule-based for another and knowledge-based for a third. The errors in my first group — slips and lapses of attention — are errors in skill-based tasks. Those errors in my second group — mistakes — that are due to poor training are mainly errors in knowledge-based tasks; those that are due to poor instructions are mainly errors in rule-based tasks. Violations can be any of these types of error.

Chapter 7 describes some of the attempts that have been made to quantify the probability of human error. However, these methods apply only to the first sort of error. We can estimate — roughly — the probability that someone will have a moment's aberration and forget to open a valve, or open the wrong valve, but we cannot estimate the probability that he will make a mistake because the training or instructions are poor, because he lacks the necessary physical or mental ability, or because he has a 'couldn't care less' attitude. Each of these factors can contribute from 0 to 100% to the probability of failure. All we can do is assume that they will continue in the future at the same rate as in the past, unless there is evidence of change. People often assume that these errors have been eliminated by selection, training, instructions and monitoring, but this is not always true.

Chapters 8–12 examine some further accidents due to human failing, but classified somewhat differently:

- accidents that could be prevented by better design;
- accidents that could be prevented by better construction;
- accidents that could be prevented by better maintenance;
- accidents that could be prevented by better methods of operation;
- accidents in computer-controlled plants.

Finally Chapter 13 looks at legal views and at the question of personal responsibility. If we try to prevent errors by changing the work situation, does this mean that people who make errors are entitled to say, 'It is your fault for putting me in a situation where I was able to make a mistake?'

1.4 Two simple examples

The various types of human error may be illustrated by considering a simple everyday error: forgetting to push in the choke on a car when the engine has warmed up (a rare error today when most cars have automatic choke). There are several possible reasons for the error:

- It may be due to a lapse of attention. This is the most likely reason and similar errors are discussed in Chapter 2.

- With an inexperienced driver the error may be due to a lack of training or instruction; he or she may not have been told what to do or when to do it. Similar errors are discussed in Chapter 3.
- The error may be due to a lack of physical or mental ability — unlikely in this case. Such errors are discussed in Chapter 4.
- The error may be due to the fact that the driver cannot be bothered — the hired car syndrome. Similar errors are discussed in Chapter 5.
- If I employ a driver, I may leave everything to him, take no interest in the way he treats the car and fail to investigate the reasons why my engines wear out so quickly. These management errors are discussed in Chapter 6.

If the error is due to lack of training or instruction then we can provide better training or instruction. There is not much we can do in the other cases except change the work situation — that is, provide a warning light or alarm or an automatic choke (now usual). The latter adds to the cost and provides something else to go wrong and in some such cases it might be better to accept the occasional error. Examples of situations in which automatic equipment is not necessarily better than an operator are given in Chapter 7.

As another example of the five sorts of error, consider spelling errors:

- If I type *opne* or *thsi* it is probably a slip, overlooked when I checked my typing. Telling me to be more careful will not prevent these errors. Using the spell-checking tool on my word processor will prevent many of them.
- If I type *recieve* or *seperate*, it is probably due to ignorance. If I type *wieght* it may be the result of following a rule that is wrong: 'I before e, except after c'. Better training or instructions might prevent the errors but persuading me to use the spell-checking tool would be cheaper and more effective.
- If I write *Llanfairpwllgwyngyllgogerychwyrndrobwllllantysiliogogogoch* (a village in Wales), it is probably because it is beyond my mental capacity to spell it correctly from memory. (Can you spot the error? Answer on page 10.)
- If I type *thru* or *gray*, it is probably due to a deliberate decision to use the American spelling. To prevent the errors someone will have to persuade me to use the English spelling.
- If I work for a company, perhaps the managers take little or no interest in the standard of spelling. They may urge me to do better but do not realize that they could do more, such as encouraging me to use the spell-checking tool.

However, the spell-checking tool can be misused and can introduce errors. The following, copied from various sources, probably occurred because someone misspelled a word, either through ignorance or as the result of a slip, and then accepted whatever the computer offered, on the assumption that the all-powerful computer could not possibly be wrong. They illustrate the way computers (and other tools) can be misused when we do not understand their limitations.

- *Tumble dryer, vertically unused.* (From a classified advertisement.)
- *The opera is in three parts. The first is an appeal to density.* (From a concert programme.)
- *Concrete arched lentils.* (From a newspaper report.)
- *Bedsit with separate fatalities.* (From a classified advertisement.)
- *Classic old brass and copper geezer.* (From a classified advertisement.)
- *Should you have any queers please contact this office.* (From a letter from a building society.)
- *A wonderful conversation of a period house to create four homes.* (From an advertisement.)
- *A red box containing assaulted tools worth £100 was stolen* ... (From a newspaper report.)

1.5 Accident investigation

The output from an accident investigation may tell us much about the culture of the company where it occurred. If it has a blame culture, people are defensive and not very forthcoming. It is difficult to find out what has happened and the true causes may never be found. In addition, when actions are discussed, the design engineer, for example, will try to show that the accident could not have been prevented by better design. People from other departments will behave similarly. In a blame-free environment, however, the design engineer will suggest ways in which better design (or changes in the design procedure, such as more use of Hazop or inherently safer design) could prevent a repetition. Each person will consider what he or she might have done to prevent the accident or make it less likely.

If we wish to find out why accidents occurred and find ways of preventing them happening again, we need a sympathetic attitude towards people who have committed so-called violations, have made slips, had lapses of attention or not learnt what their training was supposed to teach them. It is a small price and is worth paying.

This applies to everyone. We should not blame the workman who for one reason or another failed to take action that could have prevented an accident. We should ask why he did not take that action and make recommendations accordingly. (Possible reasons are discussed in the following chapters.)

Similarly we should not blame the senior manager who failed to take action that could have prevented an accident, but ask why. The reason is usually not wickedness but failure to foresee the actions he might have taken (see Chapter 6 and Sections 13.3–13.5, page 227).

Unfortunately a desire to blame someone for misfortunes seems to be deeply engrained in human nature. For example, amongst the Australian Aborigines, rites were performed often at the grave or exposure platform of the dead to discover the person to be blamed for the 'murder'. Since death was not considered a natural event a cause for it was always sought in the evil intentions of someone else, usually a member of another local group[4].

We are much the same and are ready to believe that every accident is someone's fault, especially the person at the 'sharp end'. In 1998 a plane, flying low in an Alpine Valley, broke the cable of a cable railway, causing many deaths. At first the pilot was blamed for flying too low but the inquiry showed that his altimeter was faulty, so he was much lower than he thought, and that the cable railway was not marked on his map. Nevertheless, television pictures showed relatives of those killed expressing their outrage at his acquittal. As with every accident, many people had an opportunity to prevent it. Why was the altimeter faulty? Was it poor design unsuitable for the application, or poor maintenance? Why was the railway not shown on the map? Was an old edition in use? And so on.

1.6 A story

The following story illustrates the theme of this book. A man went into a tailor's shop for a ready-made suit. He tried on most of the stock without finding one that fitted him. Finally, in exasperation, the tailor said, 'I'm sorry, sir, I can't fit you; you're the wrong shape.'

Should we as engineers expect people to change their (physical or mental) shapes so that they fit into the plants and procedures we have designed or should we design plants and procedures to fit people?

We know that they cannot change their physical shape. If a man cannot reach a valve we do not tell him to try harder or grow taller. We provide a step, move the valve or remove the need for the valve. But some people expect others to change their mental shape and never have slips or lapses of attention. This is as difficult as growing taller. We should instead change the design or method of working to reduce the opportunities for error.

1.7 Research on human error

The Third Report of the UK Advisory Committee on Major Hazards[2] and many other reports recommend research on human reliability and make some suggestions. While there is undoubtedly much we would like to know, lack of knowledge is not the main problem. The main problem is that we do not use the knowledge we have. Accidents, with a few exceptions, are not caused by lack of knowledge, but by a failure to use the knowledge that is available. This book is an attempt, in a small way, to contribute towards accident reduction by reminding readers of facts they probably know well and reinforcing them by describing accidents which have occurred, in part, as a result of ignoring these facts.

Table 1.1 summarizes the four types of accident and the corresponding actions needed. They are discussed in more detail in the following chapters.

Table 1.1 Types of human error and the action required

Error type	Action required
Mistakes — Does not know what to do	Better training and instructions / CHAOS
Violations — Decides not to do it	Persuasion / CHAOS
Mismatches — Unable to do it	CHAOS
Slips and lapses of attention	CHAOS

CHAOS = Change Hardware And / Or Software

References in Chapter 1

1. Kletz, T.A., 1984, *Plant/Operations Progress*, 3(4): 210.
2. Advisory Committee on Major Hazards, 1984, *Third Report, The Control of Major Hazards*, Appendix 13 (HMSO, London, UK).
3. Kletz, T.A., 1998, *Process Plants: A Handbook for Inherently Safer Design* (Taylor and Francis, Philadelphia, USA).
4. Australian National Commission for UNESCO, 1973, *Australian Aboriginal Culture*, page 44.

The correct spelling of the village in Wales (see page 7) is:
Llanfairpwllgwyngyllgogerychwyrndrobwllllantysiliogogogoch

Accidents caused by simple slips

2

'*I haven't got a memory, only a forgettory.*'
Small boy quoted on Tuesday Call, BBC Radio 4,
6 December 1977

'*It is a profoundly erroneous truism, repeated by all
books, and by eminent people when they are making
speeches, that we should cultivate the habit of thinking
what we are doing. The precise opposite is the case.
Civilisation advances by extending the number of
important operations which we can perform without
thinking about them.*'
A.N. Whitehead

2.1 Introduction

This chapter describes some accidents which occurred because someone forgot
to carry out a simple, routine task such as closing or opening a valve, or carried
it out wrongly — that is, he closed or opened the wrong valve. The intention
was correct; he knew what to do, had done it many times before, was capable of
doing it and intended to do it and do it correctly, but he had a moment's
aberration.

Such slips or lapses of attention are similar to those of everyday life and
cannot be prevented by exhortation, punishment or further training. We must
either accept an occasional error — probabilities are discussed in Chapter 7
— or remove the opportunity for error by a change in the work situation —
that is, by changing the design or method of working. Alternatively we can,
in some cases, provide opportunities for people to observe and correct their
errors or we can provide protection against the consequences of error.

Note that errors of the type discussed in this chapter — errors in
skill-based behaviour — occur not in spite of the fact that the man who
makes the error is well-trained but because he is well-trained. Routine tasks
are given to the lower levels of the brain and are not continuously monitored
by the conscious mind. We could never get through the day if everything we
did required our full attention, so we put ourselves on auto-pilot. When the
normal pattern or programme of action is interrupted for any reason, errors
are liable to occur.

Those interested in the psychological mechanisms by which slips and lapses of attention occur should read J. Reason and C. Mycielska's book on everyday errors[1]. The psychology is the same. Here, as engineers, we are not concerned with the reasons why these errors occur, but with the fact that they do occur and that we can do little to prevent them. We should therefore accept them and design accordingly.

Note that the unconscious mind, as used here, is not the same as the Freudian unconscious, a place of preventive detention, where we hold captive thoughts that would embarrass us if they became conscious. In contrast the non-Freudian unconscious is a mechanism which does not require consciousness to do its job. It is rather like a computer operating system running quietly in the background, or an auto-pilot[25].

People who have made an absent-minded error are often told to keep their mind on the job. It is understandable that people will say this, but it is not very helpful. No-one deliberately lets their mind wander, but it is inevitable that we all do so from time to time.

Can we reduce errors by selecting people who are less error-prone? Obviously if anyone makes a vast number of errors — far more than an ordinary person would make — there is something abnormal about his work situation or he is not suitable for the job. But that is as far as it is practicable to go. If we could identify people who are slightly more forgetful than the average person — and it is doubtful if we could — and weed them out, we might find that we have weeded out the introverts and left the extroverts and those are the people who cause accidents for another reason: they are more inclined to take chances. Section 4.3 (page 85) returns to this subject.

The examples that follow illustrate some of the ways by which we can prevent or reduce slips and lapses of attention or recover from their effects. They include better displays of information, interlocks that prevent an incorrect action being performed, warnings that it has been performed, trips which prevent serious consequences and reducing stress or distraction. Behavioural safety training (see Section 5.3 on page 107) will have little if any effect as slips and lapses of attention are not deliberate. People can be encouraged, before closing a valve, to stop and ask themselves, 'Is this the right valve?', much as we pause and look both ways, before we cross the road. However, slips and lapses are most likely to occur when we are in a hurry or under stress and these are the times when we are least likely to pause.

2.2 Forgetting to open or close a valve

2.2.1 Opening equipment which has been under pressure

A suspended catalyst was removed from a process stream in a pressure filter (Figure 2.1(a)). When a batch had been filtered, the inlet valve was closed and the liquid in the filter blown out with steam. The steam supply was then isolated, the pressure blown off through the vent and the fall in pressure observed on a pressure gauge. The operator then opened the filter for cleaning. The filter door was held closed by eight radial bars which fitted into U-bolts on the filter body. To withdraw the radial bars from the U-bolts and open the door the operator had to turn a large wheel, fixed to the door. The door, with filter leaves attached, could then be withdrawn (Figure 2.1(b) on page 14). Figure 2.2 (page 14) shows two similar filters, one open and one closed.

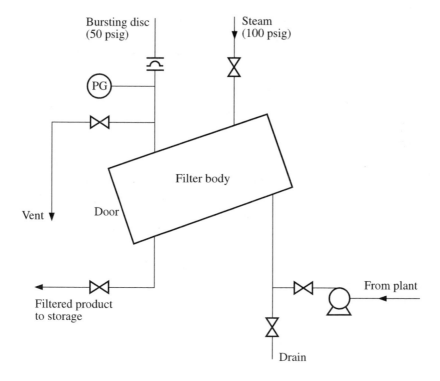

Figure 2.1(a) Filter

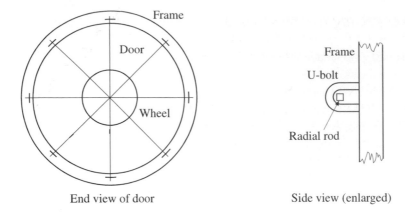

End view of door   Side view (enlarged)

Figure 2.1(b) Filter door and fastening

Figure 2.2 Two filters similar to the one involved in the accident described in Section 2.2.1. The one on the left is open and the one on the right is closed. The operator who was killed was standing in front of the wheel seen on the right. He had forgotten to vent the pressure in the filter. When he started to open the door, by turning the wheel, the door flew open.

One day an operator, a conscientious man of great experience, started to open the door before blowing off the pressure. He was standing in front of it and was crushed between the door and part of the structure. He was killed instantly.

This accident occurred some years ago and at the time it seemed reasonable to say that the accident was due to an error by the operator. It showed the need, it was said, for other operators to remain alert and to follow the operating instructions exactly. Only minor changes were made to the design.

However, we now see that in the situation described it is inevitable, sooner or later, that an operator will forget that he has not opened the vent valve and will try to open the filter while it is still under pressure. The accident was the result of the work situation, and we would recommend the following changes in the design:

(1) Whenever someone has to open up equipment which has been under pressure, using quick release devices:

(a) Interlocks should be fitted so that the vessel cannot be opened until the source of pressure is isolated and the vent valve opened (one way of doing this would be to arrange for the handles of ball valves on the steam and vent lines to project over the door handle when the steam valve is open and the vent valve closed); and

(b) The design of the door or cover should be such that it can be opened about ¼ inch (6 mm) while still capable of carrying the full pressure and a separate operation should be required to release the door fully. If the cover is released while the vessel is under pressure, this is immediately apparent and the pressure can be allowed to blow off through the gap or the door can be resealed. This is now required in the UK by the Health and Safety Executive[2,3].

(2) The pressure gauge and vent valve should be located near the door so that they are clearly visible to the operator when he is about to open the door. They were located on the floor above.

(3) The handle on the door should be modified so that it can be operated without the operator having to stand in front of the door.

Recommendations (2) and (3) were made and carried out at the time but not (1), the most important.

The accident occurred at the end of the night shift, an hour before the operator was due to start his annual holiday. His mind may not have been fully on the job; he may have been thinking of his holiday. Who can blame him? It is not very helpful to say that the accident was due to human failing.

We all have moments when for one reason or another our minds are not on the job. This is inevitable. We should list as the causes of an accident only those we can do something about. In this case the accident could have been prevented by better design of the equipment[4].

Many similar accidents have occurred when operators have had to open up equipment which has been under pressure (see Section 12.3, page 207). In contrast, every day, in every factory, equipment which has been under pressure is opened up safely for repair but this is normally done under a permit-to-work system. One person prepares the equipment and issues a permit to another person who opens it up, normally by carefully slackening bolts in case there is any pressure left inside. Safety is obtained by following procedures: the involvement of two people and the issue of a permit provides an opportunity to check that everything necessary has been done. Accidents are liable to happen when the same person prepares the equipment and opens it up and in these cases we should look for safety by design.

One design engineer, finding it difficult to install the devices recommended in (1) above, said that it was, 'reasonable to rely on the operator.' He would not have said it was reasonable to rely on the operator if a tonne weight had to be lifted; he would have installed mechanical aids. Similarly if memory tasks are too difficult we should install mechanical (or procedural) aids.

Of course, we can rely on the operator to open a valve 99 times out of 100, perhaps more; perhaps less if stress and distraction are high (see Chapter 7), but one failure in 100 or even in 1000 is far too high when we are dealing with an operation which is carried out every day and where failure can have serious results.

Many design engineers accept the arguments of this book in principle but when safety by design becomes difficult they relapse into saying, 'We shall have to rely on the operator.' They should first ask what failure rate is likely and whether that rate is tolerable (see Chapter 7).

Would a check-list reduce the chance that someone will forget to open a valve? Check-lists are useful when performing an unfamiliar task — for example, a plant start-up or shutdown which occurs only once per year. It is unrealistic to expect people to use them when carrying out a task which is carried out every day or every few days. The operator knows exactly what to do and sees no need for a check-list. If the manager insists that one is used, and that each step is ticked as it is completed, the list will be completed at the end of the shift.

When chokes on cars were manual, we all forgot occasionally to push them in when the engines got hot but we would not have agreed to complete a check-list every time we started our cars.

It is true that aircraft pilots go through a check-list at every take-off, but the number of checks to be made is large and the consequences of failure are serious.

2.2.2 Preparation and maintenance by same person

A crane operator saw sparks coming from the crane structure. He turned off the circuit breaker and reported the incident to his supervisor who sent for an electrician. The electrician found that a cable had been damaged by a support bracket. He asked a crane maintenance worker to make the necessary repairs and reminded him to lock out the circuit breaker. The maintenance worker, not realizing that the breaker was already in the 'off' position, turned it to the 'on' position before hanging his tag on it. While working on the crane he received second and third degree burns on his right hand.

The report said that the root cause was personnel error, inattention to detail, failure to confirm that the circuit was de-energized. However, this was an immediate cause, not a root cause. The root cause was a poor method of working.

As in the filter incident in Section 2.2.1, incidents are liable to occur when the same person prepares the equipment and carries out the maintenance. It would have been far better if the electrician had isolated the circuit breaker, checked that the circuit was dead and signed a permit-to-work to confirm that he had done so.

It is not clear from the report whether the circuit was locked out or just tagged out. It should have been locked out and it should have been impossible to lock it open.

The report[26] is a good example of the way managers, in writing a report, comment on the failings of the injured person but fail to see that their procedures could be improved. The incident occurred in the US; if it had occurred in the UK, the Health and Safety Executive would, I think, have commented on the poor permit system.

2.2.3 Emptying a vessel

Figure 2.3 (page 18) shows the boiler of a batch distillation column. After a batch was complete, the residue was discharged to a residue tank through a drain valve which was operated electrically from the control room. To reduce the chance that the valve might be opened at the wrong time, a key was required to operate the valve, and indicator lights on the panel showed whether it was open or shut.

One day the operator, while charging the still, noticed that the level was falling instead of rising and then realized that he had forgotten to close the

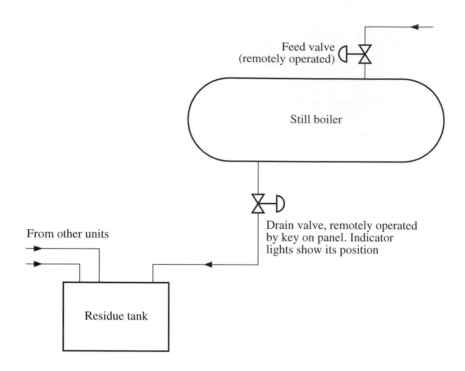

Figure 2.3 Premature discharge of the boiler caused a reaction to occur in the residue tank

drain valve after emptying the previous batch. A quantity of feed passed to the residue tank where it reacted violently with the residues.

The operator pointed out that the key was small and easily overlooked and that the indicator lights were not easily visible when the sun was shining. As an immediate measure a large metal tag was fitted to the key. (This was something that any operator could have done at any time.) Later the drain and feed valves were interlocked so that only one of them could be open at a time.

A similar incident occurred on the reactor shown in Figure 2.4. When the reaction was complete the pressure fell and the product was discharged into the product tank. To prevent the discharge valve being opened at the wrong time, it was interlocked with the pressure in the reactor so that it could not be opened until the pressure had fallen below a gauge pressure of 0.3 bar. A drain valve was provided for use after the reactor was washed out with water between batches.

A batch failed to react and it was decided to vent off the gas. When the gauge pressure fell below 0.3 bar, the discharge valve opened automatically

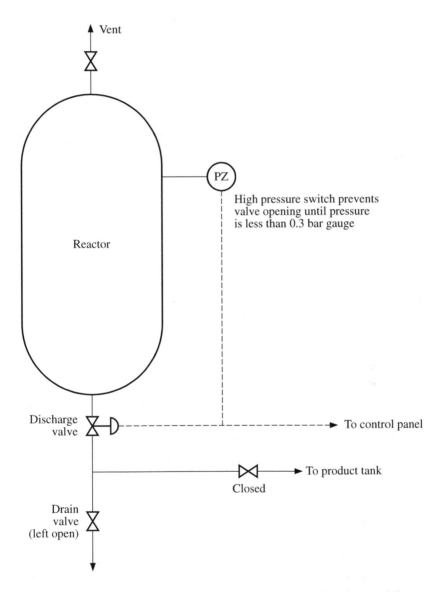

Figure 2.4 Arrangement of valves on a batch reactor. The drain valve was left
in the open position. When the pressure in the reactor fell below 0.3 bar gauge, the
discharge valve opened and the contents of the reactor were discharged.

and, as the drain valve had been left open, the contents of the reactor were discharged into the working area. Fortunately, though they were flammable, they did not catch fire. The remote actuator on the discharge valve had been left in the open position and as soon as the pressure in the reactor fell the interlock allowed the valve to open.

Again, it is too simplistic to say that the accident was the result of an error by the operator who left the actuator on the discharge valve in the open position. The accident could have been prevented by a better designed protective system.

The weakness in the design of the protective system could have been foreseen by a hazard and operability study.

2.2.4 Emptying a pump

A pump had to be drained when not in use and the drain valve left open as the liquid inside it gave off gas on standing and the pressure would damage the seal. Before starting up the pump, the drain valve had to be closed.

One day a young operator, in a hurry to start up the spare pump and prevent interruption to production, forgot to close the drain valve although he knew that he should, had done it before and an instruction to do so was included in the plant operating instructions. The liquid came out of the open drain valve and burnt him chemically on the leg. The accident was said, in the report, to be due to human failing and the operator was told to be more careful.

However, changes were made to make the accident less likely and protect the operator from the consequences of error. The drain line was moved so that the drainings were still visible but less likely to splash the operator and a notice was placed near the pump to remind the operators to close the drain valve before starting the pump. After a while the notice will have become part of the scene and may not be noticed. A better solution would have been to fit a small relief valve to the pump. Interlocking the drain valve and the starter would be possible but expensive and complex and probably not justified.

According to the accident report, ' ... it was carelessness which led him to open liquors to the pump with the vent open ... Labels will be provided to remind operators that certain pumps may be left with open vents. Process supervisors will be told to ensure that all operators and particularly those with limited experience, are fully aware of the elementary precautions they are expected to take in carrying out their duties.' These comments written by the plant manager (the equivalent of a supervisor in the US) suggest that he was rather reluctant to make any changes. The accident occurred some years ago and today most managers would be more willing to change the design. We can imagine the comments that the other people involved might have made at the time. The foreman would probably have been blunter than the manager and said the accident was entirely due to carelessness. The injured man might have said, 'It's easy to forget there is something special about a particular pump when you are busy and trying to do everything at once.' The shop steward would have said, 'Blame the working man as usual.' The safety officer might have suggested a relief valve instead of an open drain valve. The design engineer might have said, 'No-one told me that the drain valve on this pump had to be left open. If I had known I might have changed the design. Our usual practice is to fit relief valves only on positive pumps, not on centrifugal ones.'

2.3 Operating the wrong valve

2.3.1 Trip and alarm testing

A plant was fitted with a low pressure alarm and an independent low pressure trip, arranged as shown in Figure 2.5. There was no label on the alarm but there was a small one near the trip.

An instrument artificer was asked to carry out the routine test of the alarm. The procedure, well known to him, was to isolate the alarm from the plant, open the vent to blow off the pressure and note the reading at which the alarm operated.

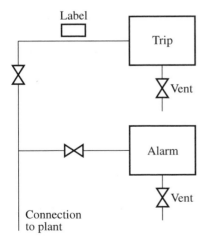

Figure 2.5 Arrangement of trip and alarm connections

By mistake he isolated and vented the trip. When he opened the vent valve, the pressure in the trip fell and the plant was automatically shut down. It took 36 hours to get it back to normal. It would have been of little use telling the artificer to be more careful.

To reduce the chance of a further mistake:

- provide better labels;
- put the trip and alarm further apart;
- possibly paint the trip and alarm different colours.

Though not relevant to this incident, note that the trip and alarm should be connected to the plant by separate impulse lines to reduce the chance of a common mode failure: choking of the common impulse line.

2.3.2 Isolation of equipment for maintenance

To save cost, three waste heat boilers shared a common steam drum. Each boiler had to be taken off line from time to time for cleaning (see Figure 2.6). On two occasions the wrong valve was closed (D3 instead of C2) and an on-line boiler was starved of water and over-heated. The chance of an error was increased by the lack of labelling and the arrangement of the valves — D3 was below C2. On the first occasion the damage was serious. High temperature alarms were then installed on the boilers. On the second occasion they prevented serious damage but some tubes still had to be changed. A series of interlocks were then installed so that a unit has to be shut down before a key can be removed; this key is needed to isolate the corresponding valves on the steam drum.

A better design, used on later plants, is to have a separate steam drum for each waste heat boiler (or group of boilers if several can be taken off line together). There is then no need for valves between the boiler and the steam drum. This is more expensive but simpler and free from opportunities for error.

Note that we do not grudge spending money on complexity but are reluctant to spend it on simplicity.

During a discussion on this accident a design engineer remarked that designers have enough problems arranging complex pipework in the space available without having to worry whether or not corresponding valves are in line with each other. This may be true but such simple things produce error-prone situations.

Figure 2.7(a) (page 24) illustrates another accident involving valves that were out-of-line. A series of parallel pipes rose up the wall of a room, went part way across overhead and then came down the centre of the room. There was a horizontal double bend in the overhead section of each pipe. As a result the pipes going up and the pipes coming down were out of line.

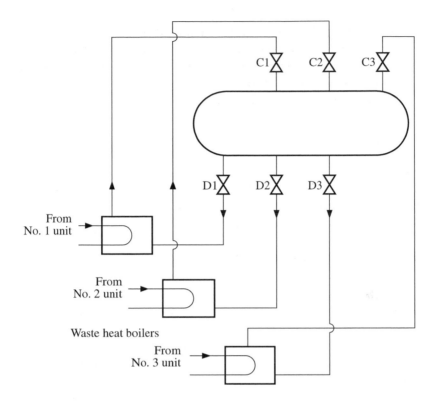

Figure 2.6 Waste heat boilers sharing a common steam drum

A valve in a pipe in the middle of the room had to be changed. One isolation valve was beneath it. The other isolation valve was on the wall. The man who isolated the valves overlooked the bends overhead and closed the valve on the wall that was in line with the valve that had to be changed. This valve was in the wrong line. When the topwork on the valve was unbolted the pressure of the gas in the line caused the topwork to fly off and imbed itself in a wall.

Figure 2.7(a) (page 24) shows the pipes as seen from above with the ceiling removed. It is easy to see which valve should have been closed. Standing amongst the pipes the correct valve is less obvious. Note that both the valves closed were the third from the end of the row as an extra pipe on the wall led elsewhere.

Colour-coding of the corresponding valves could have prevented the accident (see Figure 2.7(b) on page 24). The report did not mention this but recommended various changes in procedures. A common failing is to look

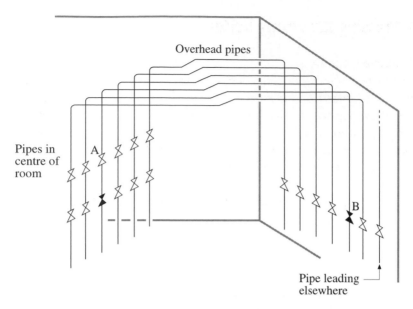

Figure 2.7(a) Valve A had to be changed. The operator closed the valve below it. To complete the isolation, he intended to close the valve on the other side of the room in the pipe leading to valve A. He overlooked the double bends overhead and closed valve B, the one opposite valve A. Both of the valves that were closed were the third from the end of their row. The bends in the overhead pipes are in the horizontal plane.

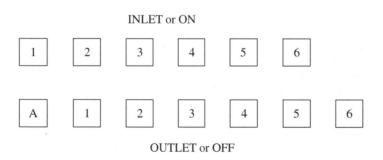

Figure 2.7(b) Valves or switches arranged like this lead to errors. If the valves or switches cannot be moved, use one colour for the pair labelled 1, another colour for the pair labelled 2, and so on.

for changes to procedures first; to consider changes in design only when changes in procedures are not possible; and to consider ways of removing the hazard rather than controlling it only as a last resort. This is the wrong order (see Section 8.7, last item, page 165).

2.4 Pressing the wrong button

2.4.1 Beverage machines

Many beverage vending machines are fitted with a panel such as that shown in Figure 2.8. I found that on about 1 occasion in 50 when I used these machines I pressed the wrong button and got the wrong drink. Obviously the consequences were trivial and not worth worrying about, but suppose a similar panel was used to charge a batch reactor or fill a container with product for sale. Pressing the wrong button might result in a runaway or unwanted reaction or a customer complaint.

It is therefore worth studying the factors that influence the probability of error and ways of reducing it.

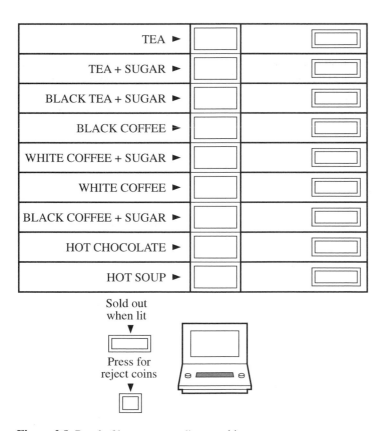

Figure 2.8 Panel of beverage vending machine

This example of human error is of interest because many of the uncertainties that are present in other examples do not apply. The errors were not due (I hope) to lack of training or instructions, or to lack of physical or mental ability. They were certainly not due to lack of motivation because when I got the wrong drink I had to drink it (being too mean or short of change to throw it away and try again). I knew what to do, was able to do it and wanted to do it, but nevertheless made an occasional error.

My error rate was increased by a certain amount of stress and distraction. (The machines are in the corridor.) It is shown in Section 7.5 (page 141) that in a situation free from stress and distraction the error rate would probably be about 3 in 1000.

If we wished to reduce the error rate we could:

- Place the machine in a place where there is less distraction; stress is harder to remove.
- Redesign the panel. We could put the buttons further apart.
- Better still, separate the choice of drink (tea, coffee, chocolate, soup) from the choice of milk (yes or no) and sugar (yes or no).

If this did not give an acceptable error rate we might have to consider attaching a microprocessor which could be told the name of the product or customer and would then allow only certain combinations of constituents. Alternatively it might display the instructions on a screen and the operator would then have to confirm that they were correct.

I moved to a new building where the beverage machines were of a different type, shown in Figure 2.9(a). I started to drink lemon tea obtained by pressing the isolated central button — and I felt sure that I would make no more mistakes. Nevertheless, after a few weeks, I did get the wrong drink. I used a machine in a different part of the building and when I pressed the centre button I got hot chocolate. The panel was arranged as shown in Figure 2.9(b).

The labelling was quite clear, but I was so used to pressing the centre button that I did not pause to read the label.

A situation like this sets a trap for the operator as surely as a hole in the ground outside the control room door. Obviously it does not matter if I get the wrong drink but a similar mistake on a plant could be serious. If there are two panels in the control room and they are slightly different, we should not blame the operator if he makes an error. The panels should be identical or entirely different. If they have to be slightly different — because of a different function — then a striking notice or change of colour is needed to draw attention to the difference.

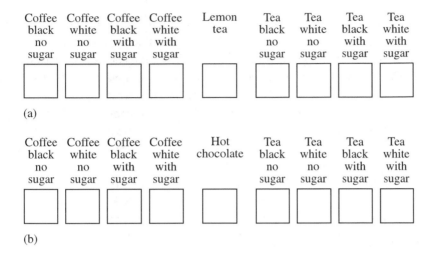

Figure 2.9 Panels of two similar vending machines

Two identical units shared a common control room. The two panels were arranged as mirror images of each other. This led to numerous errors.

There is an extensive literature on the ergonomics of control room layout.

2.4.2 Overhead cranes

Figure 2.10 (page 28) shows the buttons on the control unit for an overhead crane.

Operators become very skilled in their use and are able to press two or three buttons at a time so that they seem to be playing the controls like a concertina. Nevertheless occasional errors occur.

However, in this case the operator sees the load move the wrong way and can usually reverse the movement before there is an accident. Nevertheless we must expect an occasional 'bump'.

2.4.3 Charging a reactor

A similar mistake to the one with the beverage machine caused a serious fire in which several men were killed.

Two reactors, Nos 4 and 6, on a batch plant were shut down for maintenance. The work on No. 4 was completed and the foreman asked an operator to open the feed valve to No. 4. The valve was electrically operated and the operator went to the panel, shown in Figure 2.11 (page 29), but by mistake pressed the button controlling the inlet valve to No. 6 reactor. This reactor

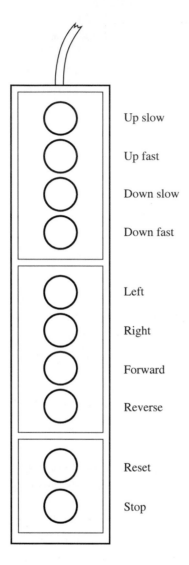

Up slow

Up fast

Down slow

Down fast

Left

Right

Forward

Reverse

Reset

Stop

Figure 2.10 Control unit for a travelling overhead crane

was still under repair. The valve opened; flammable gas came out and caught fire. The company concerned said, 'What can we do to prevent men making errors like this?' The answer is that we cannot prevent men making such errors, though we can make them less likely by putting the buttons further apart, providing better labelling and so on, but errors will still occur, particularly if the operator is under stress (see Figure 14.33, page 250).

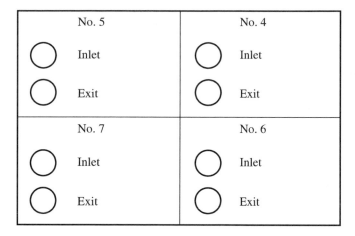

Figure 2.11 Arrangement of operating buttons for the inlet and exit valves on a group of batch reactors

We should never tolerate a situation in which such a simple slip has such serious consequences. The valves on reactors under maintenance should have been defused and locked off and the inlet and exit lines should have been slip-plated.

The operator was not to blame for the accident. He made the sort of error that everyone makes from time to time. The accident could have been prevented by a better method of working, by better management.

2.4.4 Shutting down equipment

A row of seven furnaces was arranged as shown in Figure 2.12(a) (page 30). The buttons for isolating the fuel to the furnaces were arranged as shown in Figure 2.12(b) (page 30).

An operator was asked to close the fuel valve on No. 5 furnace. He pressed the wrong button and isolated the fuel to A furnace. He realized that he had to isolate the fuel to the furnace on the extreme left so he went to the button on the extreme left.

2.5 Failures to notice

The failures discussed so far have been failures to carry out an action (such as forgetting to close a valve) or carrying it out incorrectly (closing the wrong valve).

Accidents can occur because someone fails to notice the signal for an action. We should not tell people to 'Wake up' or 'Pay more attention', but should make the signal more prominent.

The furnaces are arranged as shown below:

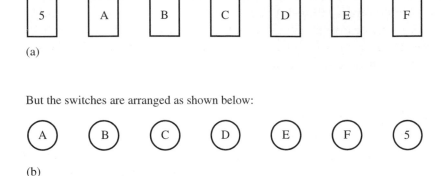

(a)

But the switches are arranged as shown below:

(b)

Figure 2.12 Arrangement of furnaces and control panel

For example, a company fitted small discs to every ladder, giving the date on which they were due for inspection. They were fitted to the top rungs.

Someone pointed out that they were more likely to be seen if they were fitted to the sixth rung from the bottom! (see Figure 14.31, page 249).

An operator charged a reactor with a chemical from a drum that looked like the drums he normally used. Unfortunately he did not notice or check the label and there was a runaway reaction. The cover of the reactor hit the ceiling, 2 m above. The report on the explosion blamed inadequate operator training, inadequate procedures and poor supervision but did not point out that if drums of different chemicals look alike then sooner or later someone, under stress or distraction, will have a momentary lapse of attention and use the wrong drum[21].

While checking a stock of dust filters for respirators, a man found that some were the wrong type and unsuitable for use. He checked the records and found that four similar ones had been issued and used. It was then found that the person who ordered the filters had left out one letter from the part number. The filters looked similar to those normally used and the storekeeper did not check the part number[27].

Reference numbers should be chosen so that a simple error (such as leaving out a letter or number, quoting the wrong one, or interchanging a pair) cannot produce a valid result.

I had a similar experience when I telephoned an insurance company to ask about progress on an accident claim. I quoted the claim number (19 letters and numbers) but the man I spoke to typed MH instead of HM when entering it into his computer. As a result he asked me to telephone another office. I repeated the number (including HM) but many of the claims they dealt with included the letters MH and they heard what they expected to hear, MH. We spoke at cross-purposes. They did not like to say that they hadn't a clue what I was talking about and brushed me off with remarks such as, 'We will keep you informed when we get more information.'

2.6 Wrong connections

Figure 2.13 shows the simple apparatus devised in 1867, in the early days of anaesthetics, to mix chloroform vapour with air and deliver it to the patient. If it was connected up the wrong way round liquid chloroform was blown into the patient with results that could be fatal. Redesigning the apparatus so that the two pipes could not be interchanged was easy; all that was needed were different types of connection or different sizes of pipe. Persuading doctors to use the new design was more difficult and the old design was still killing people in 1928. Doctors believed that highly skilled professional men would not make such a simple error but as we have seen everyone can make slips occur however well-motivated and well-trained, in fact, only when well-trained[28].

Do not assume that chemical engineers would not make similar errors. In 1989, in a polyethylene plant in Texas, a leak of ethylene exploded, killing 23

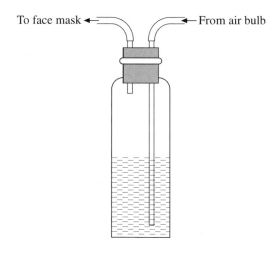

To face mask ◄— ◄— From air bulb

Figure 2.13 An early chloroform dispenser. It was easy to connect it up the wrong way round and blow liquid chloroform into the patient.

people. The leak occurred because a line was opened for repair while the air-operated valve isolating it from the rest of the plant was open. It was open because the two compressed air lines, one to open the valve and one to close it, had identical couplings, and they had been interchanged. As well as this slip there was also a violation, a decision (authorized at a senior level) not to follow the normal company rules and industry practice which required a blind flange or double isolation valve (Figure 2.14)[29] (see also Section 5.2.5, page 105).

The operating staff may have been unable to influence the design but they could have painted the two hoses and the corresponding connections different colours.

This is hardly a new idea. In the middle of the 19th century the Festiniog Railway in North Wales had a passing loop halfway along its 14-mile single-track line. Before entering each half of the line the engine driver had to possess the train staff for that section. In 1864 a railway inspector wrote, 'I have however recommended that the train staffs should be painted and figured differently for the two divisions of the line.'[30]

2.7 Errors in calculations

A batch reaction took place in the presence of an inorganic salt which acted as a buffer to control the pH. If it was not controlled, a violent exothermic side-reaction occurred. Each batch was 'tailor-made' for the particular purpose for which the product was required and the weights of the raw materials required were calculated from their compositions and the product specification.

As the result of an error in calculation, only 60% of the buffer required was added. There was a runaway reaction and the reactor exploded. Following the explosion there was a review of the adequacy of the protective system and many extra trips were added. On the modified plant, loss of agitation, high temperature or low pH resulted in the reaction being automatically aborted by addition of water. A level switch in the water tank prevented operation of the charge pump if the level was low.

In addition, the amount of buffer added was increased to twice that theoretically necessary. However, errors in calculation were still possible, as well as errors in the quantities of reactants added, and so there was a need for the protective instrumentation.

An error by a designer resulted in the supports for a small tank being too light. When the tank was filled with water for a construction pressure test, it fell to the ground, unfortunately fracturing an oil line and causing a fire which killed a man.

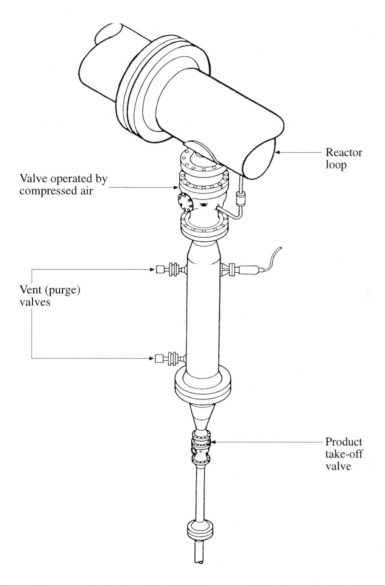

Figure 2.14 Product settling leg on a polyethylene plant. The top valve was normally open and the product take-off valve normally closed.

In another case an experienced foreman said that the supports for a new pipebridge were too far apart. The manager (myself) said, 'What you mean is that they are further apart than in the past.' 'New methods of calculation,' I suggested, 'resulted in a cheaper design.'

After a flanged joint on a pipeline on the bridge had leaked it was found that there had been an error in calculation, the pipebridge had sagged and an extra support had to be added.

Despite these two incidents, calculation errors by designers are comparatively rare and I do not suggest any changes to normal design procedures. But anything that looks odd after construction should be checked. *What does not look right, may not be right.*

As the result of a calculation error, a newborn baby was killed by an overdose of a drug. A doctor, working under stress, gave the baby 10 micrograms of a drug, per kg of body weight, instead of 1 microgram/kg. The baby's weight was reported in pounds and the error occurred in converting them to kg. An inquiry found the doctors involved not guilty of serious professional misconduct.

There are obvious ways of improving the work situation: metric units should be used throughout and labels on the drug containers could quote typical doses for patients of various weights.

The newspaper that reported the incident got the dose out by a factor of thousand; perhaps it was a slip or perhaps the reporter did not know the difference between micrograms and milligrams[31].

2.8 Other medical errors

In a long article a United States surgeon has described some of the errors that he and other doctors have made[32]. He writes:

'... *all* doctors make terrible mistakes ... [a] study found that nearly four percent of hospital patients suffered complications from treatment which prolonged their hospital stay or resulted in disability or death, and that two-thirds of such complications were due to errors in care ... It was estimated that, nationwide, 120,000 patients die each year at least partially as a result of such errors in care.

'If error were due to a subset of dangerous doctors, you might expect malpractice cases to be concentrated among a small group, but in fact they follow a uniform bell-shaped distribution. Most surgeons are sued at least once in the course of their careers. ... The important question isn't how to keep bad physicians from harming patients; it's how to keep good physicians from harming patients.'

The author goes on to suggest ways of reducing medical errors. Many hospitals hold weekly meetings at which doctors can talk candidly about their errors. Medical equipment, he says, is rife with latent errors. Cardiac defibrillators, for example, have no standard design.

'The doctor is often only the final actor in a chain of events that set him or her up to fail. Error experts, therefore, believe that it's the process, not the individuals in it, which requires closer examination and correction. In a sense they want to industrialize medicine. And they can already claim one success story: the speciality of anesthesiology, which has adopted their precepts and seen extraordinary results.'

For several months nurses in a hospital in South Africa found a dead patient in the same bed every Friday morning. Checks on the air-conditioning system and a search for infection produced no clues. Finally, the cause was discovered: every Friday morning a cleaner unplugged the life support system and plugged in her floor polisher. When she had finished she replaced the plug on the life-support system but by now the patient was dead. She did not hear any screams above the noise made by the floor polisher[33].

Section 12.5 (page 213) describes two medical errors involving computers.

2.9 Railways

By considering accidents due to human error in other industries we can check the validity of our conclusion — that occasional slips and lapses of attention are inevitable and that we must either accept them or change the work situation, but that it is little use exhorting people to take more care. Because we are not involved, and do not have to rethink our designs or modify our plants, we may see the conclusion more clearly.

Railways have been in existence for a long time, railway accidents are well documented in the official reports by the Railway Inspectorate and in a number of books[5–8,18–20, 34, 35], and they provide examples illustrating all the principles of accident investigation, not just those concerned with human error. Those interested in industrial safety will find that the study of railway accidents is an enjoyable way of increasing their knowledge of accident prevention.

2.9.1 Signalmen's errors

A signalman's error led to Britain's worst railway accident, at Quintinshill just north of the Scottish border on the London-Glasgow line, in 1915, when 226 people were killed, most of them soldiers[9].

Figure 2.15 (page 36) shows the layout of the railway lines. Lines to London are called up lines; lines from London are called down lines.

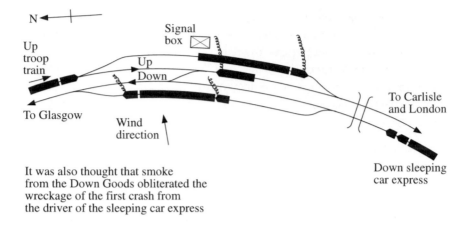

It was also thought that smoke
from the Down Goods obliterated the
wreckage of the first crash from
the driver of the sleeping car express

Figure 2.15 Layout of lines at Quintinshill, the scene of Britain's worst
railway accident

The two loop lines were occupied by goods trains and so a slow
north-bound passenger train was backed on to the up line in order to let a
sleeping car express come past. The signalman, who had just come on duty,
had had a lift on the slow train and had jumped off the foot-plate as it was
backing on to the up line. He could see the slow train through the signalbox
window. Nevertheless, he completely forgot about it and accepted a
south-bound troop train which ran into the slow train. A minute or so later the
north-bound express train ran into the wreckage. The wooden coaches of the
troop train caught fire and many of those who survived the first impact were
burned to death.

The accident occurred because *the signalman forgot that there was a train
on the up line*, though he could see it from his window and had just got off it,
and accepted another train. A contributory cause was the failure of the
signalman who had just gone off duty to inform the signalman in the next
signalbox that the line was blocked and to put a reminder collar on the signal
lever.

One signalman had a lapse of memory — and obviously it was not
deliberate.

The other signalman was taking short cuts — omitting to carry out jobs
which he may have regarded as unnecessary.

2.9.2 What should we do?

As in the other incidents discussed in this chapter, there are three ways of preventing similar incidents happening again:

(1) Change the hardware.

(2) Persuade the operators to be more careful.

(3) Accept the occasional accident (perhaps taking action to minimize the consequences).

(1) Changing the hardware was, in this case possible, but expensive. The presence of a train on a line can complete a 'track circuit' which prevents the signal being cleared. At the time, track-circuiting was just coming into operation and the inspector who conducted the official enquiry wrote that Quintinshill, because of its simple layout, would be one of the last places where track-circuiting would be introduced. It was not, in fact, installed there until the electrification of the London-Glasgow line many years later; many branch lines are still not track-circuited.

(2) Both signalmen were sent to prison — it was war-time and soldiers had been killed.

It is doubtful if prison, or the threat of it, will prevent anyone forgetting that there is a train outside their signalbox, especially if they have just got off it.

Prison, or the threat of it, might prevent people taking short cuts, but a better way is management supervision. Did anyone check that the rules were followed? It would be surprising if the accident occurred on the first occasion on which a collar had been left off or another signalman not informed (see Chapter 5).

(3) In practice, since prison sentences were probably ineffective, society accepted that sooner or later other similar accidents will occur. They have done, though fortunately with much less serious consequences.

Sometimes accepting an occasional accident is the right solution, though we do not like to admit it. So we tell people to be more careful if the accident is trivial, punish them if the accident is serious and pretend we have done something to prevent the accident recurring. In fact, the accidents arise out of the work situation and, if we cannot accept an occasional accident, we should change the work situation.

Although a similar accident could happen today on the many miles of UK branch lines that are not track-circuited, the consequences would be less serious as modern all-steel coaches and modern couplings withstand accidents and fires much better than those in use in 1915.

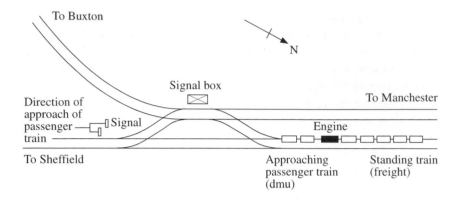

Figure 2.16 Track layout at Chinley North Junction. The signalman forgot that the freight train was standing on the track and allowed the passenger train to approach.

A similar accident to Quintinshill occurred in 1979 on British Railways, though the consequences were less serious. A signalman forgot there was a freight train standing on the wrong line and accepted another train (Figure 2.16). The track-circuiting prevented him releasing the signal so he assumed there was a fault in the track-circuiting and displayed a green hand signal to the driver of the oncoming train. He did not even check the illuminated diagram in his signalbox which would have shown the standing train[10].

As at Quintinshill, there was an irregularity in the signalman's procedures. He should have gone down to the track to give the green hand signal, not displayed it from the signalbox. Had he done so, he might have seen the standing train.

This incident shows how difficult it is to design protective equipment which is proof against all human errors. If signalmen are not permitted to use green hand lamps, what do they do when track-circuiting and signals go out of order? The incident is also an example of a mind-set, discussed later (Section 4.4, page 88). Once we have come to a conclusion we close our minds to further evidence and do not carry out the simplest checks.

A change since 1915 is that there was no suggestion that the signalman should be punished. Instead the official report wonders if his domestic worries made an error more likely.

2.9.3 Drivers' errors – signals passed at danger (SPADs)

Many accidents have occurred because drivers passed signals at danger. At Aisgill in the Pennines in 1913, 14 people were killed and the driver, who survived, was imprisoned[6]. He was distracted by problems with his engine and the Inquiry criticized the management for the strictness with which they punished delay and incorrect treatment of engines. As a result of this policy drivers were more afraid of being late than of having an accident. At Harrow in 1952, 112 people, including the driver, were killed. At Moorgate in 1975, 42 people, including the driver, were killed[12]. The driver is the person most at risk and many other drivers have been killed in this way, so they have no incentive to break the rules, but accidents still occur. Many are described in official reports.

In March 1989 a British Railways train driver passed a caution signal without slowing down. He did not brake until he saw a red signal and his train crashed into the rear of the train in front. Five people were killed and 88 injured. In September 1990, in a throwback to the 19th century, the driver was prosecuted for manslaughter, convicted and sentenced to six months' imprisonment, reduced to four months on appeal, of which two were actually served.

The official report on the accident[39] pointed out that the signal involved had been passed at red on several previous occasions. After the accident a 'repeater' was installed to improve its visibility. The managers who did not provide the equipment that could have prevented the accident were not, of course, unconcerned about safety. At worst they did not understand the nature of human error; at best they may have decided that more lives would be saved if the railway's resources were spent in other ways.

Following a similar accident in 1996, another driver was prosecuted for manslaughter but acquitted[40].

Davis[11] has analysed a number of cases in detail and has shown that while a few of the drivers were clearly unsuited to the job, the majority were perfectly normal men with many years' experience who had a moment's aberration.

As with signalmen, we should accept an occasional error (the probability is discussed in Section 7.8, page 147) — this may be acceptable on little-used branch lines — or install a form of automatic train protection. On the UK Automatic Warning System (AWS) a hooter sounds when a driver approaches a caution (amber or double amber) or a stop (red) signal and if he does not cancel the hooter, the brakes are applied automatically. It is possible for the driver to cancel the hooter but take no further action. This is the correct procedure if the speed of the train is already low! On busy lines drivers are constantly passing signals at amber and cancelling the alarm can become almost automatic.

This became increasingly common during the 1980s and in 1986 the Chief Inspector of Railways, in his Annual Report for 1985, asked British Railways to consider a form of Automatic Train Protection (ATP) which could not be cancelled by the driver and which would automatically apply the brakes to prevent the train passing a stop signal. Even at this date some railway officials did not see the need to do anything more to prevent driver errors. The cost was high and the saying 'The driver is paid to obey signals' was often heard[22]. Following further accidents, in 1989 British Railways agreed to implement a system over a period of ten years[23]. However, the cost turned out to be excessive and the Government refused to authorize it. Ten years later, no changes had been made and Railtrack, the now privatized successor to British Railways, agreed to adopt a simpler but less effective system than that originally proposed. In the Train Protection and Warning System (TPWS), as it is called, the speed of a train is measured. If it is approaching a danger signal, a buffer stop or a speed restriction too rapidly, the brakes are applied automatically. They are also applied if a train passes a signal at danger, whatever the speed[36]. (Unofficially TPWS is known as ATP-Lite.) In August 1999 the Government approved regulations requiring TPWS to be fitted to 40% of the track by the end of 2003[37].

In March 2001 a Public Inquiry[43] recommended that ATP should be installed on all major railway lines, to prevent on average two deaths per year at a cost of some tens of millions of pounds per year, perhaps more. Road engineers could save far more lives with this money. Nevertheless, ATP was recommended as SPADs have the potential to kill some tens of people in a single accident — 31 were killed at Paddington in 1999 — and as public opinion is particularly averse to railway accidents[38]. Many people will see this recommendation as democracy in action. Others may regard it as giving the most to those who shout the loudest.

The psychological mechanisms that may have caused the driver to act as he did are discussed by Reason and Mycielska[13]. From an engineering viewpoint, however, it is sufficient to realize that for one reason or another there is a significant chance (see Section 7.8, page 147) of a driver error and we must either accept the occasional error or prevent it by design (see Section 13.3, page 227).

For many years the concentration on human error diverted attention from what could be done by better engineering, such as:

- Improving the visibility of signals, especially those where SPADs have occurred more than once (about two-thirds of all SPADs).
- Redesigning track layouts so as to reduce conflicting movements, where one traing has to cross the path of another. 90% of the accidents due to SPADs have occurred as a result of such movements.
- Providing drivers with diagrams of little-used routes through complex junctions and taking them through on simulators.

2.10 Other industries

2.10.1 Aircraft

Many aircraft have crashed because the pilots pulled a lever the wrong way[14]. For example, most modern jets are fitted with ground spoilers, flat metal plates hinged to the upper surface of each wing, which are raised *after* touch-down to reduce lift. They must not be raised *before* touch-down or the aircraft will drop suddenly.

On the DC-8 the pilot could either: (1) *Lift* a lever before touch-down to arm the spoilers (they would then lift automatically after touch-down); or (2) Wait until after touch-down and *pull* the same lever.

One day a pilot *pulled* the lever before touch-down. Result: 109 people killed.

The accident was not the fault of the pilot. It was the result of bad design. It was inevitable that sooner or later someone would move the lever the wrong way.

The reaction of the US Federal Aviation Administration was to suggest putting a notice in each cockpit alongside the spoiler lever saying 'Deployment in Flight Prohibited'. They might just as well have put up a notice saying 'Do Not Crash this Plane'.

The manufacturer of the DC-8, McDonnell Douglas, realized the notice was useless but wanted to do nothing. After two, perhaps three, more planes had crashed in the same way they agreed to fit interlocks to prevent the ground spoilers being raised before touch-down.

Many other aircraft accidents are blamed on pilot error when they could have been prevented by better design. Hurst writes:

'Some 60% of all accidents involve major factors which can be dismissed as "pilot error". This sort of diagnosis gives a ... feeling of self-righteousness to those who work on the ground; but I want to state categorically that I do not believe in pilot error as a major cause of accidents. There are, it is true, a very few rare cases where it seems clear that the pilot wilfully ignored proper procedures and got himself into a situation which led to an accident. But this sort of thing perhaps accounts for one or two per cent of accidents — not 60%. Pilot error accidents occur, not because they have been sloppy, careless, or wilfully disobedient, but because we on the ground have laid booby traps for them, into which they have finally fallen.'[15]

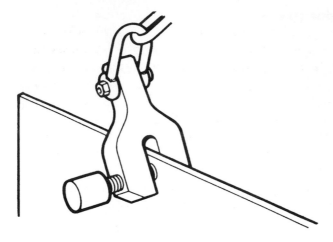

(a) The screw must be tightened to the right extent.

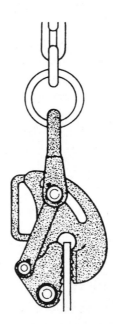

(b) This design is not dependent on someone tightening a screw to the correct extent.

Figure 2.17 Two methods of attaching a metal plate to a hoist

2.10.2 Roads

Motorists, like engine drivers, also can pass red signals as the result of a temporary lapse of attention but in this case the police have no option but to assume that the action was deliberate, or many people would pass them regularly.

An 'Accident Spotlight' issued jointly by a police force and a local radio station gave a diagram of a road junction where seven injury-causing accidents occurred in one year and then said, 'Principal cause: road user error'.

There is no reason to believe that road users behave any worse at this junction than at any other and little hope that they will change their ways. Drawing attention to the number of accidents that occur at the junction may persuade a few people to approach it more cautiously but it would be better to redesign the junction. That however is expensive. It is cheaper to blame road users.

2.10.3 Mechanical handling

A plate fell from a clamp while being lifted because the screw holding it in position had not been tightened sufficiently (Figure 2.17(a)). The operator knew what to do, but for some reason did not do it. The accident was put down as due to human failing and the operator was told to be more careful in future.

It would have been better to use a type of clamp such as that shown in Figure 2.17(b) which is not dependent for correct operation on someone tightening it up to the full extent. The accident could be prevented by better design.

2.11 Everyday life (and typing)

A man whose telephone number is 224 4555 gets frequent calls for a doctor whose number is 224 5555. It seems that people look up the doctor's number, say 'four fives' to themselves and then key a four, followed by several fives. People may be under some stress when telephoning the doctor and this may make errors more likely.

I am indebted to my former secretary, Eileen Turner, for the following incident.

In an unusually houseproud mood, she cleaned the bedroom before going early to bed one night. She woke the next morning at six o'clock and, finding she couldn't get back to sleep, decided to get up and wash her hair.

After showering, brushing her teeth and washing her hair, she went into the living room, where after a few minutes, she noticed that the time by the

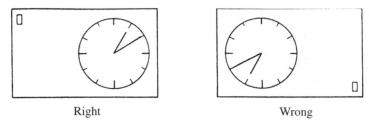

Right   Wrong

Figure 2.18 Two ways of putting a clock down

rather old clock there was ten past one (Figure 2.18). The clock had obviously had its day and was going haywire but Eileen went to the bedroom to check. On first glance the time was now twenty to seven but closer examination showed that the clock was upside down! (Figure 2.18).

To prevent a similar incident happening again she could change the hardware — modify the top of the clock so that it could not be put down upside down — or change the software — that is, give up dusting!

The writer, Bill Bryson, admits that he has gone out to buy some tobacco and post a letter, bought the tobacco first and then posted it. He also reports that a bank robber who covered his face with a balaclava forgot to remove his identity badge, with his name on, from his clothing[41].

A contractor was hired to demolish a row of dilapidated cowsheds. He sent a driver with an excavator to do the job. Instead of the cowsheds on one side of the road, the driver demolished an historic 18th century farmhouse on the other side[42].

Fire brigade officers (not in the UK) were provided with a 50-page booklet which listed in three columns the names of chemicals they might come across, their United Nations numbers and the fire-fighting agent to be used. The typist left the first line in the third column blank, and moved all the other entries in that column one line down, as shown in Table 2.1. The booklets were in use for many months before the error was spotted.

Table 2.1 The top line in the third column was accidentally left blank and all the entries in the column were moved one row down

United Nations number	Name	Fire-fighting agent
1001	Abcd	
1002	Efgh	Water
1003	Ijkl	Dry powder
1004	Mnop	Foam

The following two 'human failing' accidents occurred during demonstration lectures.

'The experimenter demonstrated the power of nitric acid to the subject by throwing a penny into it. The penny, of course, was completely disintegrated ... While the subject's view of the bowl of acid was cut off by the experimenter, an assistant substituted for it a like-sized bowl of ... water ... The hypnotized subject was then ordered to throw the dish of nitric acid (in actual fact, of course, innocuous water) over the assistant who was present in the same room. Under these conditions it was possible to induce, under hypnosis, various subjects to throw what they considered to be an extremely dangerous acid into the face of a human being ... Actually, in this particular experiment the person in charge made what he calls a "most regrettable mistake in technique" by forgetting to change the nitric acid to the innocuous dish of water, so that in one case the assistant had real nitric acid thrown over him.'[16]

'A deplorable accident has taken place at the Grenoble Lycée. The professor of chemistry was lecturing on salts of mercury, and had by his side a glass full of a mercurial solution. In a moment of distraction he emptied it, believing he was drinking a glass of eau sucrée. The unfortunate lecturer died almost immediately.'[17]

The first accident could be prevented by not indulging in such a foolish experiment, which can so easily and obviously go wrong, and the second by using different containers for drinks and laboratory chemicals.

Many other slips, mostly trivial in their consequences, are described by Reason and Mycielska[1].

2.12 Fatigue

Fatigue may make slips, and perhaps errors of other sorts, more likely. In the Clapham Junction railway accident (see Section 6.5, page 118) a contributory factor to a slip was 'the blunting of the sharp edge of close attention' due to working seven days per week without a day off. The man involved was not, however, physically tired.

The report on a chemical plant accident (according to press reports a flare stack was filled with oil which overflowed, producing flames several hundred feet high) said, 'It is clear that in this case operating teams overlooked things that should not have been overlooked, and misinterpreted instrument readings. A major influence over the behaviour of the operating teams was their tiredness and frustration.' A trade union leader is quoted as saying that the

45

management team members were more tired than the operators as the managers were working 12-hour shifts[24].

Obviously we should design work schedules that do not produce excessive fatigue, but inevitably some people will be tired from time to time as the result of factors that have nothing to do with work. When we can we should design plants and methods of working so that fatigue (like other causes of error) does not have serious effects on safety, output and efficiency.

References in Chapter 2

1. Reason, J. and Mycielska, C., 1982, *Absent-minded? The Psychology of Mental Lapses and Everyday Errors* (Prentice-Hall, Englewood Cliffs, New Jersey, USA).
2. *Technical Data Note 46, Safety at Quick-opening and Other Doors of Autoclaves*, 1974 (Factory Inspectorate, London, UK).
3. *Guidance Note FM/4, High Temperature Dyeing Machines*, 1976 (HMSO, London, UK).
4. Kletz, T.A., 1980, *Loss Prevention*, 13: 1.
5. Rolt, R.T.C., 1987, *Red for Danger*, 4th edition (David and Charles, Newton Abbott, UK).
6. Schneider, A. and Masé, A., 1968, *Railway Accidents of Great Britain and Europe* (David and Charles, Newton Abbot, UK).
7. Hamilton, J.A.B., 1981, *Trains to Nowhere*, 2nd edition (Allen and Unwin, London, UK).
8. Gerard, M. and Hamilton, J.A.B., 1984, *Rails to Disaster* (Allen and Unwin, London, UK).
9. Hamilton, J.A.B., 1969, *Britain's Greatest Rail Disaster* (Allen and Unwin, London, UK).
10. Olver, P.M., 1981, *Railway Accident. Report on the Collision that Occurred on 14th February 1979 at Chinley North Junction in the London Midland Region of British Railways* (HMSO, London, UK).
11. Davis, D.R., 1966, *Ergonomics*, 9: 211.
12. McNaughton, I.K.A., 1976, *Railway Accident. Report on the Accident that Occurred on 18th February 1975 at Moorgate Station on the Northern Line, London Transport Railways* (HMSO, London, UK).
13. As Reference 1, page 204.
14. Eddy, P., Potter, E. and Page, B., 1976, *Destination Disaster* (Hart-Davis and MacGibbon, London, UK).
15. Hurst, R. and L.R. (editors), 1982, *Pilot Error*, 2nd edition (Aaronson, New York, USA).
16. Eysenck, H.J., 1957, *Sense and Nonsense in Psychology* (Penguin Books, London, UK).

17. *Nature*, 18 March 1880, quoted in *Nature*, 20 March 1980, page 216.
18. Nock, O.S. and Cooper, B.K., 1982, *Historic Railway Disasters*, 4th edition (Allen and Unwin, London, UK).
19. Hall, S., 1987, *Danger Signals* (Ian Allen, London, UK).
20. Reid, R.C., 1968, *Train Wrecks* (Bonanza Books, New York, USA).
21. *Loss Prevention Bulletin*, 1989, No. 090, page 29.
22. As Reference 19, page 126.
23. Hidden, A. (Chairman), 1989, *Investigation into the Clapham Junction Railway Accident*, Sections 14.27–14.31 and 15.8–15.12 (HMSO, London, UK).
24. *Evening Gazette* (Middlesbrough), 24 and 25 August 1987.
25. Miller, J. in Silvers, R.B (editor), 1997, *Hidden Histories of Science*, page 1 (Granta Books, London, UK and New York Review of Books, New York, USA).
26. *Operating Experience Weekly Summary*, 1998, No. 98–52, page 3 (Office of Nuclear and Facility Safety, US Department of Energy, Washington, DC, USA).
27. *Operating Experience Weekly Summary*, 1999, No. 99–17, page 1 (Office of Nuclear and Facility Safety, US Department of Energy, Washington, DC, USA).
28. Sykes, W.S., 1960, *Essays on the First Hundred Years of Anaesthesia*, Chapter 1 (Churchill Livingstone, Edinburgh, UK).
29. *The Phillips 66 Company Houston Chemical Complex Explosion and Fire*, 1990 (US Department of Labor, Washington, DC, USA).
30. Tyler, W.H., 1864, in Public Record Office file MT 29 25, page 676, quoted by Jarvis, P., 1999, *Ffestiniog Railway Magazine*, Spring, No. 164, page 329.
31. Weaver, M., *Daily Telegraph*, 6 March 1999; Deacock, A.R., *ibid.*, 8 March 1999; Davies, C., *ibid.*, 21 April 1999; Poole, O., *ibid.*, 14 April 1999.
32. Gawande, A., 1999, *The New Yorker*, 1 February, page 20.
33. *Cape Times*, 13 June 1996, quoted in *IRR News*, August 1997 (Institute for Risk Research, Toronto, Canada).
34. Hall, S., 1989, *Danger on the Line* (Ian Allen, London, UK).
35. Hall, S., 1996, *British Railway Accidents* (Ian Allen, London, UK).
36. *Modern Railways*, June 1999, page 417.
37. *Modern Railways*, September 1999, page 659.
38. Evans, A., *Modern Railways*, September 1999, page 638.
39. *Report on the Collision that Occurred on 4th March 1989 at Purley in the Southern Region of British Railways*, 1990 (HMSO, London, UK).
40. Marsh, A., *Daily Telegraph*, 23 March 1998.
41. Bryson, B., 1998, *Notes from a Big Country*, pages 104 and 294 (Doubleday, London, UK).
42. Pile, S., undated, *The Return of Heroic Failures*, page 18 (Secker & Warburg, London, UK).
43. Uff, J. and Cullen, W.D., 2001, *The Joint Inquiry into Train Protection Systems* (HSE Books, Sudbury, UK).

Accidents that could be prevented by better training or instructions

3

'I am under orders myself, with soldiers under me. I say to one, Go! and he goes; to another, Come here! and he comes; and to my servant, Do this! and he does it.'
A centurion in St. Matthew, Chapter 8, Verse 9
(New English Bible)

'The painful task of thinking belongs to me. You need only obey implicitly without question.'
Admiral Sir George Rodney, to his captains, 1708[24]

3.1 Introduction

These quotes may have been true at one time, they may still be true in some parts of the world, but they are no longer true in Western industry and I doubt if they are still true in the armed services. Tasks have changed and people's expectations have changed.

According to a book on the war-time Home Guard, the part-time defence force[25], 'As General Pownall knew from bitter experience, the Home Guard could be coaxed and persuaded to act in a certain way, but it could not be treated like the army where orders were simply orders.'

Training as used here includes education (giving people an understanding of the technology and of their duties) and also includes teaching them skills such as the ability to diagnose faults and work out the action required, while instructions tell them what they should and should not do. Oversimplifying, we can say that training should prevent errors in knowledge-based and skill-based behaviour while instructions should prevent errors in rule-based behaviour. These errors are called mistakes to distinguish them from the slips and lapses of attention discussed in Chapter 2.

If a task is very simple, instructions may be sufficient, but in the process industries most tasks are no longer simple. We cannot lay down detailed instructions to cover every contingency. Instead we have to allow people judgement and discretion and we need to give them the skills and knowledge they need to exercise that judgement and discretion. We should give them the skill and knowledge needed to diagnose what is wrong and decide what action to take. We shall look at some accidents that occurred because people were not adequately trained.

A biologist who has studied the ability of insects to learn writes that this ability has evolved when the environment is too unpredictable for fixed, inborn behaviour patterns to be appropriate. The individual, if it is to survive, has to learn how to adjust its behaviour in response to changing circumstances. Many insects — indeed, most of those investigated — have been shown to be capable learners[39].

In the same way, process plants are too unpredictable to be always operated by fixed rules. Operators have to learn how to adjust their behaviour in response to changing circumstances.

People's expectations have changed. They are no longer content to do what they are told just because the boss tells them to. Instead they want to be convinced that it is the right thing to do. We need to explain our rules and procedures to those who will have to carry them out, and discuss them with them, so that we can understand and overcome their difficulties.

Of course, this argument must not be carried too far. Nine of your staff may agree on a course of action. The tenth man may never agree. He may have to be told, 'We have heard and understand your views but the rest of us have agreed on a different course of action. Please get on with it.'

The man who thinks he knows but does not is more dangerous than the man who realizes he is ignorant. Reference 20 describes accidents which occurred as the result of incorrect beliefs. Always check to make sure that training and instructions have been understood. The message received may not be the same as the one transmitted. Sometimes no message is received. If operators are given a list of flash points or occupational exposure standards, do they know what they mean or how they should use them?

The examples that follow illustrate some of the ways we can prevent mistakes — that is, accidents that occur because people do not know the correct action to take:

- The obvious way is to improve the training or instructions, but before doing so we should try to simplify the job or redesign the work situation so as to reduce opportunities for error (most of this chapter).
- Train contractors as well as employees (Section 3.3.4, page 56).
- Tell people about changes in equipment or procedures (Section 3.3.5, page 57).
- Explain the designer's intentions and the results of not following them (Sections 3.3.7 and 3.3.8, page 61).
- Avoid contradictions (Section 3.5, page 65).

- Make people aware of the limitations of their knowledge (Section 3.6, page 66).
- Check that instructions are easy to read, readily available and written to help the reader rather than protect the writer. Explain and discuss new instructions with the people who are expected to follow them (Section 3.7, page 67).
- Train people to diagnose faults and act accordingly. Instructions cannot cover all eventualities (Section 3.8, page 72).
- Include the lessons of past accidents in training (Sections 3.3.1 and 3.3.3, pages 53 and 54).

Discussions are better than lectures or writing as people have an opportunity for feedback and they remember more.

Sometimes accidents have occurred because of a lack of sophisticated training, as at Three Mile Island; sometimes because of a lack of basic training.

3.2 Three Mile Island

The accident at Three Mile Island nuclear power station in 1979 had many causes and many lessons can be drawn from it[1,2] but some of the most important ones are concerned with the human factors. In particular, the training the operators received had not equipped them to deal with the events that occurred.

To understand these I must briefly describe the design of the pressurized water reactor of the type used at Three Mile Island, and the events of 20 March 1979.

Figure 3.1 shows a very simplified drawing. Heat generated in the core by radioactive fission is removed by pumping primary water round and round it. This water is kept under pressure to prevent it boiling (hence it is called a pressurized water reactor, to distinguish it from a boiling water reactor). The primary water gives up its heat to the secondary water which does boil. The resulting steam then drives a turbine, before being condensed and the condensate recycled. All the radioactive materials, including the primary water, are enclosed in a containment building so that they will not escape if there is a leak.

The trouble started when a choke occurred in a resin polisher unit, which removes impurities from the secondary water. To try to clear the choke the operators used instrument air. Unfortunately it was at a lower pressure. As a result water got into the instrument air lines and the turbine tripped.

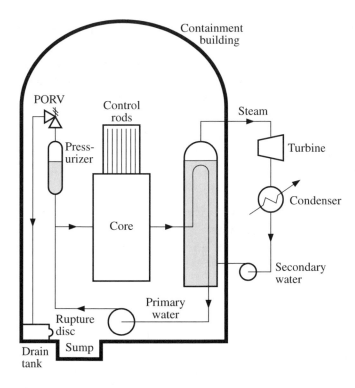

Figure 3.1 A pressurized water reactor — simplified

This stopped the heat being removed from the radioactive core. The production of fission heat stopped automatically because silver rods which absorb neutrons dropped down into the core and stopped the radioactive fission. However, heat was still produced by radioactive decay at about 6% of the full load, and this caused the primary water to boil. The pilot operated relief valve (PORV) lifted and the make-up pumps started up automatically to replace the water which had evaporated from the primary circuit. Unfortunately the PORV stuck open. The operators did not realize this because a light on the panel told them it was shut. However, this light was not operated by the valve position but by the signal to the valve. The operators had not been told this or had forgotten.

Several other readings should have suggested to the operators that the PORV was stuck open and that the primary water was boiling. However, they chose to believe the PORV light and ignore the other readings, partly because they did not really understand how the temperature and pressure in the primary circuit depended on each other, and partly because their instructions and training had emphasized that it was dangerous to allow the primary circuit to get too full of water.

The operators thought the PORV was shut. Conditions were clearly wrong and their training had emphasized the danger of adding too much water. They therefore shut down the make-up water pumps.

Note that the only action taken by the operators made matters worse. If they had done nothing, the system would have cooled down safely on its own. With the make-up water isolated, however, the level in the primary circuit fell and damage occurred.

The training of the operators was deficient in three major respects:

(1) It did not give them an understanding of the phenomena taking place in the primary water circuit; in particular, as already stated, they did not understand how the pressure and temperature were related.

(2) It did not tell them how to recognize a small loss of water — though it covered a major loss such as would occur if there was a break in the primary circuit — and what action to take if this occurred.

(3) It did not train them in the skills of diagnosis. We cannot foresee everything that will go wrong and write instructions accordingly — though what did go wrong should have been foreseen — and so we need to train operators to diagnose previously unforeseen events.

One successful method of doing so has been described by Duncan and co-workers[3]. The operator is shown a mock-up display or panel on which various readings are shown. From them he has to diagnose the fault and say what action he would take. The problems gradually increase in difficulty. The operator learns how to handle all foreseeable problems and acquires general skills which will help him handle unforeseen problems.

Better training will make a repeat of Three Mile Island less likely but a better solution, in the long term, is to develop designs of nuclear reactor that are more user-friendly — that is, less affected by human error (of any sort) or equipment failure. Gas-cooled reactors and fast reactors (the fast refers to the speed of the neutrons) are friendlier than water-cooled reactors and a number of other inherently safer designs have been proposed[2].

The accident at Chernobyl nuclear power station was also due to human error but in this case the primary error was a failure to follow instructions, though lack of understanding may have played a part (see Section 5.1.1, page 101).

3.3 Other accidents that could be prevented by relatively sophisticated training

The training required may not seem very sophisticated until we compare the incidents described in Section 3.4 (page 62).

3.3.1 Re-starting a stirrer

A number of accidents have occurred because an operator found that a stirrer (or circulation pump) had stopped and switched it on again. Reactants mixed suddenly with violent results.

For example, an acidic effluent was neutralized with a chalk slurry in a tank. The operator realized that the effuent going to drain was too acidic. On looking round, he found that the stirrer had stopped. He switched it on again. The acid and the chalk reacted violently, blew off the manhole cover on the tank and lifted the bolted lid. No-one was injured.

A similar incident had occurred on the same plant about four years earlier. An instruction was then issued detailing the action to be taken if the stirrer stopped. The operator had not seen the instruction or, if he had seen it, he had forgotten it. No copy of it could be found on the plant.

It is difficult to prevent accidents such as this by a change in design. An alarm to indicate that the stirrer had stopped might help and the manhole on the tank should be replaced by a hinged lid which will lift when the pressure rises, but it will still be necessary to train the operators and maintain the instructions. Initial training is insufficient. Regular refreshers are necessary. It is also useful to supplement training with a plant 'black book', a folder of reports on incidents that have occurred. It should be compulsory reading for newcomers and it should be kept in the control room so that others can dip into it in odd moments. It should not be cluttered up with reports on tripping accidents and other trivia, but should contain reports on incidents of technical interest, both from the plant and other similar plants.

Instructions, like labels, are a sort of protective system and like all protective systems they should be checked regularly to see that they are complete and maintained as necessary.

3.3.2 Clearing choked lines

Several incidents have occurred because people did not appreciate the power of gases or liquids under pressure and used them to clear chokes.

For example, high pressure water wash equipment was being used to clear a choked line. Part of the line was cleared successfully, but one section remained choked so the operators decided to connect the high pressure water directly to the pipe. As the pressure of the water was 100 bar (it can get as high as 650 bar) and as the pipe was designed for only about 10 bar, it is not surprising that two joints blew. Instead of suspecting that something might be wrong, the operators had the joints remade and tried again. This time a valve broke.

Everyone should know the safe working pressure of their equipment and should never connect up a source of higher pressure without proper authorization by a professional engineer who should first check that the relief system is adequate.

On another occasion gas at a gauge pressure of 3 bar — which does not seem very high — was used to clear a choke in a 2 inch pipeline. The plug of solid was moved along with such force that when it hit a slip-plate it made it concave. Calculation, neglecting friction, showed that if the plug weighed 0.5 kg and it moved 15 m, then its exit velocity would be 500 km/hour!

An instrument mechanic was trying to free, with compressed air, a sphere which was stuck inside the pig chamber of a meter prover. Instead of securing the chamber door properly he fixed it by inserting a metal rod — a wheel dog — into the top lugs. When the gauge pressure reached 7 bar the door flew off and the sphere travelled 230 m before coming to rest, hitting various objects on the way[4].

In all these cases it is clear that the people concerned had no idea of the power of liquids and gases under pressure. Many operators find it hard to believe that a 'puff of air' can damage steel equipment.

In the incident described in Section 2.2 (page 13), the other operators on the plant found it hard to believe that a pressure of 'only 30 pounds' caused the door of the filter to fly open with such force. They wondered if a chemical explosion had occurred.

The operators did not understand the difference between force and pressure. They did not understand that a force of 30 pounds was exerted against every square inch of a door that was 3.5 feet in diameter and that the total force on the door was therefore 20 tons.

3.3.3 Failures to apply well-known knowledge

It is not, of course, sufficient to have knowledge. It is necessary to be able to apply it to real-life problems. Many people seem to find it difficult. They omit to apply the most elementary knowledge. For example, scaffolding was erected around a 75 m tall distillation column so that it could be painted. The scaffolding was erected when the column was hot and then everyone was surprised that the scaffolding became distorted when the column cooled down[5].

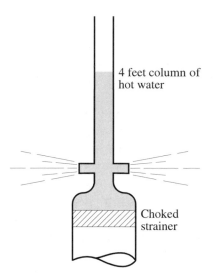

Figure 3.2 A choked strainer had to be removed. There was a column of hot water 4 feet (1.2 m) tall above the liquid. The joint above the liquid was broken to let the water run out. It came out with such force that two men several feet away were sprayed and scalded.

We all know the weight of water but many people are surprised by the pressure produced by a column of it (Figure 3.2). As a rough rule of thumb, a column of liquid x feet tall may spread x feet through a cracked joint.

Many pressure vessels have burst when exposed to fire. For example, at Feyzin in France in 1966 a 2000 m^3 sphere containing propane burst, killing 18 people and injuring many more[6–10]. The vessel had been exposed to fire for 1½ hours before it burst and during this time the fire brigade had, on the advice of the refinery staff, used the available water for cooling surrounding vessels to prevent the fire spreading. The relief valve, it was believed, would prevent the vessel bursting.

I have discussed the Feyzin fire on many occasions with groups of students and with groups of experienced operating staff and their reaction is often the same: the relief valve must have been faulty or too small or there must have been a blockage in its inlet or exit pipe. When they are assured that the relief valve was OK it may still take them some time to realize that the vessel burst because the metal got too hot and lost its strength. Below the liquid level the boiling liquid kept the metal cool, but above the liquid level the metal softened and burst.

Everyone knows that metal loses its strength when it becomes hot, but there was a failure to apply that knowledge, both by the refinery staff at the time and by the people who attended my discussions.

However, today people realize why the sphere burst more quickly than they did in the past.

Failure to apply the knowledge we have is not, of course, a problem peculiar to plant operators. One of the major problems in education is not giving knowledge to people, but persuading them to use it. Most of us keep 'book learning' and 'real life' in separate compartments and the two rarely meet. One method of helping to break down the barrier between them is by the discussion of incidents such as Feyzin. The group puzzle out why it occurred and say what they think should be done. The Institution of Chemical Engineers provides sets of notes and slides for use in this way[11].

Here is another example of failure to apply well-known knowledge because it belongs to a different compartment of life. Some empty 45-gallon drums were transported from sea level to a site at 7200 feet (2200 m). The pressure inside the drums was then 3 psi (0.2 bar) above the outside pressure. When the drums were being opened, the lid of one was forcibly ejected. We all know the pressure falls as the altitude increases but no-one applied the knowledge. Afterwards people recalled that similar incidents had occurred before and that a restraining device had been acquired to hold the lids in position while the pressure was released[26].

3.3.4 Contractors

Many accidents have occurred because contractors were not adequately trained.

For example, storage tanks are usually made with a weak seam roof, so that if the tank is overpressured the wall/roof seam will fail rather than the wall/floor seam. On occasions contractors have strengthened the wall/roof seam, not realizing that it was supposed to be left weak.

Many pipe failures have occurred because contractors failed to follow the design in detail or to do well what was left to their discretion[12]. The remedy lies in better inspection after construction, but is it also possible to give contractors' employees more training in the consequences of poor workmanship or short cuts on their part? Many of them do not realize the nature of the materials that will go through the completed pipelines and the fact that leaks may result in fires, explosions, poisoning or chemical burns. Many engineers are sceptical of the value of such training. The typical construction worker, they say, is not interested in such things. Nevertheless, it might perhaps be tried. (Construction errors are discussed further in Chapter 9.)

3.3.5 Information on change

Accidents have occurred because changes in design necessitated changes in construction or operating methods, but those concerned were not told. Our first example concerns construction.

In Melbourne, Australia in 1970, a box girder bridge, under construction across the Yarra river, collapsed during construction. The cause was not errors in design, but errors in construction: an attempt to force together components that had been badly made and did not fit.

However it is customary for construction teams to force together components that do not fit. No-one told them that with a box girder bridge — then a new type — components must not be forced together; if they do not fit they must be modified. The consulting engineers made no attempt to ensure that the contractors understood the design philosophy and that traditional methods of construction could not be used. Nor did they check the construction to see that it was carried out with sufficient care[13].

A 25 tonne telescopic jib crane was being used to remove a relief valve, weight 117 kg, from a plant. The jib length was 38 m and the maximum safe radius for this jib length is 24 m. The driver increased the radius to 31 m and the crane fell over onto the plant (Figure 3.3). Damage was slight but the plant had to be shut down and depressured while the crane was removed. The

Figure 3.3 This crane tried to lift too great a load for the jib length and elevation

crane was fitted with a safe load indicator of the type that weighs the load through the pulley on the hoist rope; it does not take into account the weight of the jib. Because of this the driver got no warning of an unsafe condition and, as he lifted the valve, the crane overturned. The driver had been driving telescopic jib cranes for several years but did not appreciate the need not to exceed the maximum jib radius. He did not realize that the crane could be manoeuvred into an unstable position without an alarm sounding.

Those responsible for training the driver had perhaps failed to realize themselves that the extra degree of freedom on a telescopic jib crane — the ability to lengthen the jib — means that it is easier to manoeuvre the crane into an unsafe position. Certainly they had failed to take this into account in the training of the driver. They had, incidentally, also failed to realize that an extra degree of freedom requires a change in the method of measuring the approach to an unsafe condition.

In the summer of 1974, a plastic gas main was laid in a street in Freemont, Nebraska, alongside a hotel, and was fixed to metal mains at each end by compression couplings. In the following winter the pipe contracted and nearly pulled itself out of the couplings. The next winter was colder and the pipe came right out. Gas leaked into the basement of the hotel and exploded. The pipe was 106 m long and contracted about 75 mm[14].

Apparently nobody told the men who installed the pipe that when plastic pipe is used in place of metal pipe, it is necessary to allow for contraction. (This might be classified as an accident due to failure to apply well-known knowledge. Everyone knows that substances contract on cooling.)

At a more prosaic level, accidents have occurred because people were not told of changes made while they were away. An effluent had to be neutralized before it left a plant. Sometimes acid had to be added, sometimes alkali. Sulphuric acid and caustic soda solution were supplied in similar plastic containers (polycrates). The acid was kept on one side of the plant and the alkali on the other. While an operator was on his days off someone decided it would be more convenient to have a container of acid and a container of alkali on each side of the plant. When the operator came back no-one told him about the change. Without checking the labels he poured some excess acid into a caustic soda container. There was a violent reaction and he was sprayed in the face. Fortunately he was wearing goggles[15] (Figure 3.4).

A small catalyst storage tank had the connections shown on the left of Figure 3.5. Valve B became blocked and so the lines were rearranged as shown on the right. An operator who had been on temporary loan to another plant returned to his normal place of work. He was asked to charge some catalyst to the plant. This used to be done by opening valve B but nobody

* *Always check labels*
* *Tell people about changes made while they were away*

Figure 3.4 A change was made while an operator was away

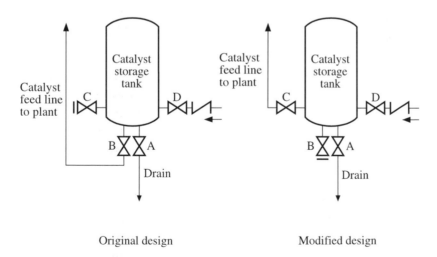

Figure 3.5 When valve B became blocked the lines were rearranged but nobody told the operator who was away at the time

59

told him about the change. When he got to the catalyst tank he noticed that something seemed to be different. The blanked valve B was obviously not the feed valve so he opened the valve alongside it, A. Several tons of an expensive catalyst drained into a sump where they reacted with rainwater and produced toxic fumes.

A better operator would have traced the lines to make quite sure before opening any valve, but proper labelling would have prevented the accident.

A common failing is not changing instructions or training after a change in organization. To quote from a report on a case of water hammer, 'operational responsibility for the laundry steam system had been assigned to a new organization. The previous organization's practice had been to drain the steam system at a known low point before re-introducing steam following an outage … The new organization was not given a turnover on how to drain the system and they tried to drain it at a location that was not the low point.' Fortunately there was no damage to the steam lines, only violent banging, but water hammer has ruptured steam lines on many other occasions[27].

It is bad practice to rely on operational methods to overcome poor design. More steam traps should have been fitted.

3.3.6 Team working

With the reduction in manpower on many plants it has become fashionable to employ 'team working'. The operators are not assigned specific tasks but each is capable of carrying out the whole range of operations and each task is done by whoever is most conveniently available at the time.

This obviously encourages efficiency, gives people a larger range of interests and sounds good, but there are snags:

- Extensive training is necessary to make sure that every operator can really carry out the full range of tasks.
- A situation in which anyone can do a job, can easily become a situation in which no-one does it. Everybody thinks someone else is doing it. A system is needed for making sure that this does not occur.

One incident occurred (in part) because a control room operator, over the loudspeaker, asked the outside operators to close a valve. There were two outside operators who worked as a team. The control room operator did not ask a particular outside man to close the valve. Both were some distance away, and each left it to the other whom he thought would be nearer.

3.3.7 The limitations of instructions

The following incidents show what can occur when a unforeseen problem arises and people have not been trained to understand the hazards and the design intention.

Aluminium powder was manufactured by spraying molten liquid through an atomizing nozzle. It was then sorted by size by conveying it pneumatically through a series of cyclones and screens. The concentration of powder was kept below the minimum concentration at which a dust explosion was possible. The powder was then transferred into road tankers and offloaded using the exhaust from the tanker's engine. The concentration of the powder could not be controlled but this did not matter as the oxygen content of the exhaust gas was low.

One day the contents of a tanker had to be recycled through the sorting section of the plant. The operators decided to empty the tanker in the usual way, using its exhaust gas. The tanker was put as near to the plant as possible but its hoses would not reach. The operators looked for an extension hose but the only one they could find had a larger bore. No problem; they stuck the tanker's hose inside the end of the larger one, filled the gap with rags and wrapped tape round the joint. They then realized that the velocity in the larger hose would be low, so to assist the transfer they inserted a compressed air line inside the joint between the two hoses. They did not see anything wrong with this as air was used for moving the powder inside the plant. They did not understand anything about explosive concentrations of either powder or oxygen. When the transfer started there was an explosion. The source of ignition was an electrostatic discharge from the unearthed metal parts of the large hose[28].

A Hazop gives the commissioning or operating manager a good understanding of the designer's intentions which he can then include in the operating instructions.

3.3.8 Ignorance of what might occur

Air filters contaminated with lithium hydride were made safe by placing them in a vat of water. The water reacted with the hydride forming lithium hydroxide and hydrogen. One day a filter contained more lithium hydride than usual. The heat developed by the reaction burnt the wooden frame of the filter, even though it was under water; some bubbles of air were probably entrapped. The hydrogen exploded, blew the lid off the vat and scattered pieces of the filter[29].

Hazard and operability studies (Hazops) force us to consider deviations such as 'More of', but simple operations such as putting a piece of discarded material in water are not usually Hazoped. This incident shows that they should be when hazardous materials are handled. Once we ask whether more lithium hydride than usual could be present we are halfway to preventing the accident. Obvious reasons are more hydride in the air being filtered or delay in changing a filter.

This incident is also an example of forgotten knowledge. At one time old filters were perforated to release trapped air before they were immersed in water. This was not in the instructions but was a 'trick of the trade', forgotten when experienced people left and were replaced (see also Section 3.7.1, page 67).

A Hazop brings out the consequences of not following the designer's intentions and these can be included in the plant instructions.

3.4 Accidents that could be prevented by elementary training

This section describes some incidents which occurred because people lacked the most elementary knowledge of their duties or the properties of the material they handled, knowledge so basic that it was considered unnecessary to impart it. What seems obvious to us may not be obvious to others.

(1) Many accidents have occurred because operators wrote down abnormal readings on the plant record sheets but did nothing about them. They did not even tell their foremen. They thought their job was just to take readings, not to respond to them. There is an example in Section 4.4.2 (page 91).

(2) After changing a chlorine cylinder, two men opened valves to make sure there were no leaks on the lines leading from the cylinder. They did not expect to find any so they did not wear breathing apparatus. Unfortunately there were some leaks, and the men were affected by the chlorine.

If we were sure there would be no leaks, there was no need to test. If there was a need to test, then leaks were possible and breathing apparatus should have been worn.

(3) A liquid was transferred into a tank by vacuum. The tank was emptied with a vacuum pump and then the pump was shut down and isolated and the liquid drawn into the tank. This was not quick enough for one shift who kept the pump running.

Someone told them that spray from the inflow would be drawn into the pump. The operators said, 'No, it doesn't; you see, all the time the pump is running there's a strong stream of "vac" running into the tank and it keeps the splashes away.'[16] It is fair to point out that this incident occurred many years ago.

(4) An operator had to empty tankers by gravity. His instructions said:

(i) Open the vent valve on the top of the wagon.

(ii) Open the drain valve.

(iii) When the tanker is empty, close the vent valve.

One day he had an idea. To save climbing onto the top of the tanker twice, he decided to carry out step (iii) before step (ii). Result: the wagon was sucked in.

The operator did not understand that when a liquid flows out of a tank it leaves an empty space behind.

(5) Before the days of self-service petrol pumps, the attendants seemed to be the most untrained of all employees, despite the hazardous nature of the material they handled. For example:

- A man got out of a mini-van just in time to prevent the attendant filling it up through the ventilator in the roof.
- A young attendant used his cigarette lighter to check the level in a road tanker.
- An attendant put petrol into the oil filler pipe of a rear-engined van. The engine caught fire when the ignition was switched on.

In this last case the attendant was watched by the driver who himself removed the oil filler cap[17].

(6) Soon after green (that is, lead-free) petrol became available, a university safety officer saw petrol dripping from a car in the university car park. As it was near the gatehouse, he asked the gateman if he had seen it and taken any action. The gateman said, 'Don't worry; it's this safer, green petrol that doesn't burn.'

(7) Maintenance workers as well as operators occasionally display the most extraordinary ignorance of what practices are or are not acceptable.

A 24 inch manhole branch on a vessel had corroded and had to be replaced. A new branch was made and the old one removed. When the new branch was offered up to the opening in the vessel it was found to be a little bit too small; the old branch was 24 inches internal diameter, while the new one was 24 inches external diameter.

The supervisor therefore decided to make a series of parallel cuts in the branch, splay them out until they fitted the hole in the vessel and weld up the gaps! Fortunately before the job was complete it came to the notice of a senior engineer and was stopped.

(8) A reaction mixture started to boil violently, rising towards the vessel manhole. An operator tried to contain it by closing the manhole lid. Fortunately

he was not successful or the vessel might have burst. As it was, the jet of liquid escaping through the manhole punched a hole in the asbestos-cement roof 4 m above[30].

Engineers as well as operators can be unaware of the power exerted by a runaway reaction. Almost all reactions go faster as the temperature rises. Nevertheless the designers of a plant came to the conclusion that a certain reaction would fade away if the temperature rose above a certain figure. When it did rise, there was an explosion and people were killed[27].

(9) Some drums of resin, normally stored in a cool location (as recommended in the material safety data sheet), were moved to a location where the temperature was at times above 45°C. Three operators who opened the drums were exposed to vapour concentrations above the short-term exposure limit quoted in the material safety data sheet. With the passage of time everyone had forgotten why the drums were kept in a cool area[31].

(10) A subcontractor, employed to remove contamination from the plant equipment, sprayed a water-based solvent onto a box containing 440-volt electrical equipment, causing serious arcing. The person who authorized the work should have warned the subcontractor. Don't assume that everybody knows what seems obvious. A year later the incident was repeated on a site that had been sent the report[32].

(11) Two operators were asked to transfer some sludge into a 60 gallon drum using a positive pump that delivered 1–1.3 gauge (15–20 psig). They left the lid on the drum to prevent splashing. When the pressure in the drum reached the delivery pressure of the pump, the transfer stopped. One of the operators decided to remove the lid. While trying to loosen the locking ring, the lid flew off and hit him on the forehead[38].

Training rather than instructions is the better way to prevent accidents such as this. It should not be necessary to tell operators not to pump with the lid fastened. Operators should be given enough understanding of pressure and its formation to make such detailed instructions unnecessary. It would be difficult for those who write instructions to think of all the situations in which pressure might be present.

(12) Software companies have considered changing the instruction 'Strike any key' because so many people have telephoned them to say that they cannot find the 'Any' key[33].

(13) An American at a car rental desk in the UK was overheard complaining that the car she had been given was very slow. She had never before driven a car

with a manual gearbox. She put the gear lever in the position marked 1 and left it there[34]. Knowledge that is obvious to many may not be obvious to people with a different background.

Other incidents are described in Chapter 9.

In a sense all these incidents and the others described in this chapter were due to human failing but it is not very helpful to say so. The incidents could have been prevented by better management, in these cases by better training and instructions, or by the other ways suggested in Section 3.1 (page 48).

3.5 Contradictory instructions

Operators sometimes receive contradictory instructions. For example, they may be told never to use wheel-dogs on stiff valves but may still be expected to operate them. One suspects that the author of the instruction wants to turn a blind eye but also wants to be able to avoid responsibility when a valve is damaged.

Operators were instructed to heat a reactant to 45°C as they added it to a reactor over a period of 60–90 minutes. They believed this to be impossible, because the heater was not powerful enough, so, without telling anybody, they decided to add it at a lower temperature and heat the material already in the reactor. This went on for a long time, and became custom and practice, until a runaway reaction occurred, with emission of toxic fumes.

Unfortunately, if operators believe that their instructions cannot be followed, they do not like to say so and instead they take what they think is the nearest equivalent action. If the manager had examined the record sheets regularly, or even occasionally, he would have noticed that his instructions were not being followed. A first-line manager should try to look at every figure on the record sheets every day, or at least several times per week.

Heavy road vehicles are carried through the Channel Tunnel between England and France on railway wagons. These wagons are fitted with supports (props) which are lowered when vehicles are being loaded and unloaded. If a prop descends at other times, an alarm sounds in the driving cab and the driver is instructed to stop the train. If a fire occurs drivers are instructed not to stop until they are out of the tunnel.

A vehicle caught fire in 1996; the fire alarm sounded and the driver continued with the journey. Soon afterwards another alarm indicated that a prop had dropped. The driver then had contradictory instructions. He decided to stop the train as a dropped prop might derail it. The prop had not dropped, the alarm was false and stopping the train greatly increased the damage to the train and to the tunnel.

The official report is not entirely clear but it seems likely that the operation of the prop alarm was the result of damage to the alarm system by the fire. It is most unlikely that a random failure occurred by coincidence during the fire[35,36]. Unfortunately no-one foresaw that two alarms, requiring different actions, might operate at the same time and that their operation could be contingent rather than coincidental.

Most contradictory instructions are, however, more subtle than those just described. A manager emphasizes the importance of completing repairs, experiments or a production run in a certain period of time. Without actually saying so he gives the impression that the normal safety procedures can be relaxed. If there is an accident he can say that he never authorized any relaxation.

Sometimes relaxation of the normal safety rules is justified. If so the manager should say so clearly in writing and accept responsibility. He should not drop hints and put his staff in a 'Heads I win, tails you lose' position (see Section 3.8, page 72).

A report from the US Congress on allegations that arms were sold to Iran and the proceeds used, illegally, to buy arms for the Nicaraguan rebels said, 'The President created, or at least tolerated, an environment where those that did know of the diversion believed with certainty that they were carrying out the President's policies.' It also said, 'If the President did not know what his national security advisers were doing, he should have.'[21] See also Sections 5.1 and 13.2 (pages 99 and 223).

3.6 Knowledge of what we don't know

The explosion at Flixborough in 1974, which killed 28 people, was due to the failure of a large (0.7 m) diameter temporary pipe which was designed and installed very quickly by men who had great practical experience and drive but did not know how to design large pipes operating at high temperature and pressure. This is understandable; the design of large, highly-stressed pipes is a specialized branch of mechanical engineering. But they did not know this and did not realize that they should have called in an expert in piping design. Instead they went ahead on their own. Their only drawing was a full-sized sketch in chalk on the workshop floor[22,23].

They could not be blamed. They lacked the professional training which would have allowed them to see when expert advice was needed. They did not know what they did not know. They were unconscious of their incompetence. The responsibility lay with the senior managers who asked them to undertake a task for which they were unqualified. The factory normally employed just one professionally-qualified mechanical engineer. He had left and his replacement had not yet arrived.

With reductions in staff, people may be asked to undertake additional duties. A control engineer, for example, may be made responsible for electrical maintenance. There will be a fully-qualified electrical engineer available for consultation, but does the control engineer know when to consult him? Similarly, if a company dispenses with in-house experts and decides to hire outside consultants as and when required, will the staff know when they ought to be consulted?

3.7 Some simple ways of improving instructions

As already stated, training gives people the knowledge and understanding they need to do their job properly, while instructions tell them the precise course of action to be followed.

In addition to the well-established methods of training, there are newer methods touched on in Sections 3.2 and 3.3.3 (pages 50 and 54). Here are some questions that should be asked about instructions.

- Are they easy to read?
- Are they written to help the reader or to protect the writer?
- Are they explained to those who will have to carry them out?
- Are they maintained?

3.7.1 Are they easy to read?

Many instructions are not. Figure 3.6 (page 68) is typical of the language and appearance of many. Figure 3.7 (page 69) shows a much better layout.

Men are remarkably good at detecting meaning in a smog of verbiage but they should not be expected to do so. Sooner or later they will fail to comprehend. With computer control, precise instructions are essential (see Chapter 12).

Table 3.1 (page 70) is an instruction (based on Reference 37) for the transfer of liquid into a batch tank. It was developed in consultation with the operators and is intended for the training of new operators. Unlike many similar instructions, it explains the reasons for the various steps. The example is rather simple but illustrates the need to explain why as well as what.

INSTRUCTION NO: WC 101

TITLE: HOW TO LAY OUT OPERATING INSTRUCTIONS
SO THAT THEY MAY BE READILY DIGESTED
BY PLANT OPERATING STAFF.

AUTHOR: EAST SECTION MANAGER

DATE: 1 DECEMBER 1996

COPIES TO: UNCLE TOM COBBLEY AND ALL

Firstly, consider whether you have considered every
eventuality so that if at any time in the future anyone
should make a mistake whilst operating one of the plants on
East Section, you will be able to point to a piece of paper
that few people know exists and no-one other than yourself
will have read or understood. Don't use one word when five
will do, be meticulous in your use of the English language
and at all times ensure that you make every endeavour to add
to the vocabulary of your operating staff by using words
with which they are unfamiliar; for example, never start
anything, always initiate it. Remember that the man reading
this has turned to the instructions in desperation, all else
having failed, and therefore this is a good time to
introduce the maximum amount of new knowledge. Don't use
words, use numbers, being careful to avoid explanations or
visual displays, which would make their meaning rapidly
clear. Make him work at it; it's a good way to learn.

Whenever possible use the instruction folder as an
initiative test; put the last numbered instruction first, do
not use any logic in the indexing system, include as much
information as possible on administration, maintenance data,
routine tests, plants which are geographically close and
training, randomly distributed throughout the folder so that
useful data are well hidden, particularly that which you need
when the lights have gone out following a power failure.

Figure 3.6 Are your plant instructions like this?

The following extract from a plant instruction shows the action a supervisor and four operators should take when the induced draught fan providing air to a row of furnaces (known as A side) stops. Compare the layout with that of Figure 3.6.

ACTION TO TAKE WHEN A SIDE ID FAN TRIPS

1 CHECK A STREAM FURNACES HAVE TRIPPED

2 ADJUST KICK-BACK ON COMPRESSORS TO PREVENT SURGING

3 REDUCE CONVERTER TEMPERATURES

4 CHECK LEVEL IN STEAM DRUMS TO PREVENT CARRY-OVER

Panel Operator

1 Shut TRCs on manual

2 Reduce feed rate to affected furnaces

3 Increase feed to Z furnace

4 Check temperature of E54 column

Furnace Operator

1 Fire up B stream and Z furnaces

2 Isolate liquid fuel to A stream furnaces

3 Change over superheater to B stream

4 Check that output from Z furnace goes to B stream

Centre Section Operator

1 Change pumps onto electric drive

2 Shut down J43 pumps

Distillation Operator

1 Isolate extraction steam on compressor

2 Change pumps onto electric drive

Figure 3.7 ... or like this?

Table 3.1 An instruction that explains the reason for each step

Instructions for charging Compound A to Batch Tank 1

Step	Details
1. Wear standard protective clothing and rubber gloves.	Standard protective clothing is hard hat, safety glasses and safety shoes.
2. Check that Tank 1 is empty.	**If tank is not empty it will overflow.**
3. Check that inlet valves to other tanks are closed.	**If valves are not closed another tank might overflow.**
4. Check that discharge valve on Tank 1 is closed.	Valve is located on far side of tank from flowmeter and is labelled.
5. Open the inlet valve on Tank 1.	Valve is located near the flowmeter and is labelled.
6. Enter charge quantity on flowmeter.	Charge quantity is shown on batch card.
7. Start the transfer using the switch on the flowmeter.	Check that the pump has started. Check the level indicator on Tank 1 is rising. If not, stop pump and check for closed valve.
8. During transfer, check for leaks.	If a leak is found, stop pump and inform supervisor.
9. When flowmeter reaches setpoint, check that flow has stopped.	If flow has not stopped, shut down pump manually.
10. Close inlet valve to Tank 1.	
11. Start stirrer on Tank 1.	
12. Charge inhibitor.	Batch card shows type and amount. **Failure to add inhibitor could cause polymerization and rise of temperature.**

3.7.2 Are they written to help the reader or to protect the writer?

It is not difficult to recognize instructions written to protect the writer. They are usually written in a legalistic language, are very long and go into excessive detail. As a result they are often not read. An instruction that covers 95% of the circumstances that might arise, and is read and understood, is better than one that tries to cover everything but is not read or not understood.

3.7.3 Are they explained to those who will have to carry them out?

On one works the instructions on the procedure to be followed and the precautions to be taken before men were allowed to enter a vessel or other confined space ran to 23 pages plus 33 pages of appendices; 56 pages in all. There were many special circumstances but even so this seems rather too long. However, when the instruction was revised it was discussed in draft with groups of supervisors and the changes pointed out. This was time-consuming for both supervisors and managers but was the only way of making sure that the supervisors understood the changes and for the managers to find out if the changes would work in practice. Most people, on receiving a 56-page document, will put it aside to read when they have time — and you know what that means. New instructions should always be discussed with those who will have to carry them out (see Section 6.5, page 118).

3.7.4 Are they maintained?

Necessary maintenance is of two sorts. First, the instructions must be kept up-to-date and, second, regular checks should be made to see that the instructions are present in the control room and in a legible condition. If too worn they should obviously be replaced. If spotlessly clean, like poetry books in libraries, they are probably never read and the reasons for this should be sought. Perhaps they are incomprehensible.

A senior manager visiting a control room should ask to see the instructions — operating as well as safety. He may be surprised how often they are out-of-date, or cannot readily be found or are spotlessly clean.

One explosion occurred because an operator followed out-of-date instructions he found in a folder in the control room.

Finally, a quotation from H.J. Sandvig:

'Operators are taught by other operators and … each time this happens something is left unsaid or untold unless specific operating instructions are provided, specific tasks are identified and written and management reviews these procedures at least annually and incorporates changes and improvements in the process.'[18]

A plant contained four reactors in parallel. Every three or four days each reactor had to be taken off line for regeneration of the catalyst. The feed inlet and exit valves were closed, the reactor was swept out with steam and then hot air was passed through it. One day a fire occurred in the reactor during regeneration. Afterwards the staff agreed that the steam purging should be carried out for longer and should be followed by tests to make sure that all the feed had been swept out. An instruction was written in the shift handover log and a handwritten note was pinned up in the control room but no change was made to the folder of typed operating instructions.

A year later an operator who had been on loan to another unit returned to his old job. He saw that the note on extra sweeping out and testing had disappeared and assumed that the instruction had been cancelled. He did not carry out any extra sweeping out or any tests. There was another, and larger, fire.

There was no system for maintaining instructions. (In addition, the content of the instructions was inadequate. Feed may have been entering the reactor through a leaking valve. The reactor should have been isolated by slip-plates or double block and bleed valves.)

Section 11.5 (page 194) describes some further incidents which could have been prevented by better instructions.

3.7.5 Other common weaknesses in instructions

- They do not correspond to the way the job is actually done.
- They contain too much or too little detail. Complex unfamiliar tasks with serious results if carried out incorrectly need step-by-step instructions. Simple familiar tasks may be tools of the trade and need no instructions at all. Different levels of detail are needed by novices and experienced people. For example, if a line has to be blown clear with compressed gas, novices should be told how long it will take and how they can tell that the line is clear. Table 3.1 (page 70) is an example of an instruction for novices.
- The reasons for the instructions are not clear (see Chapter 5).
- The boundaries of space, time and job beyond which different instructions apply are not clear.

3.8 Training or instructions?

As stated at the beginning of this chapter, training gives us an understanding of our tasks and equips us to use our discretion, while instructions tell us precisely what we should and should not do; training equips us for knowledge-based and skill-based behaviour while instructions equip us for rule-based behaviour. Which do we want?

While admitting that most jobs today require some knowledge-based behaviour, many managers feel that in some situations, especially those involving safety, rule-based behaviour is essential. In practice the rules never cover every situation and people have to use their discretion. They are then blamed either for not following the rules or for sticking rigidly to the rules when they were obviously inapplicable, another example of contradictory instructions (see Section 3.5, page 65). To try to cover every situation the rules become more and more complex, until no-one can understand them or find the time to read them. People may adopt the attitude that since they are going to be blamed in any case when things go wrong, it hardly matters whether they follow rules or not.

It would be better to recognize that people have to be given some discretion, give them the necessary training, distinguish between the rules that should never (well, hardly ever) be broken and those that on occasions may be and accept that from time to time people will make the wrong decisions.

Of course, whenever possible, exceptions to safety rules should be foreseen and authorized in advance.

The results of trying to legislate for all eventualities have been described by Barbara Tuchman, writing about naval warfare in the 18th century[24]. In the 17th century individual captains used their own tactics, and this often caused unmanageable confusion. The UK Admiralty therefore issued Fighting Instructions, which required ships to act in concert under the signalled order of the commanding officer and forbade action on personal initiative.

'In general, the result did make for greater efficiency in combat, though in particular instances … it could cause disaster by persuading a too-submissive captain to stick by the rule when crisis in a situation could better have been met by a course determined by the particular circumstances. As deviations from the rule were always reported by some disgruntled officer and tried by a court-martial, the Instructions naturally reduced, if not destroyed, initiative except when a captain of strong self-confidence would act to take advantage of the unexpected. Action of this kind was not infrequent, even though no people so much as the British preferred to stay wedded to the way things had always been done before. In allowing no room for the unexpected that lies in wait in the waywardness of men, not to mention the waywardness of winds and oceans, Fighting Instructions was a concept of military rigidity that must forever amaze the layman.'

In 1744, after the battle of Toulon, Admiral Mathews was court-martialled and dismissed for not sticking strictly to the rules. Aware of this decision, in

1757 Admiral Byng (who had been on the bench which tried Mathews) stuck to the rules at the battle of Minorca and was court-martialled and shot for not having done his utmost to relieve the garrison of Minorca. (This produced the famous saying of Voltaire, 'Dans ce paysci il est bon de tuer de temps en temps un amiral pour encourager des autres.' ['In that country they find it pays to kill an admiral from time to time to encourage the others.']) During the American War of Independence many British officers refused to accept commands, as they were afraid of being made scapegoats if anything went wrong.

Though the penalties are less, similar incidents have occurred in industry. For example, a waste heater boiler is fitted with a low water level alarm and the operators are told to shut it down at once if the alarm sounds. An operator does so; the alarm was false and production is lost. Nothing is said to the operator but he senses that everyone wishes he had not shut the plant down. Perhaps someone suggests to the foreman that this operator's talents might be better employed on another unit, where strict attention to the rules is necessary. Everyone gets the message. Next time the alarm sounds it is ignored and the boiler is damaged!

Getting the right balance between following the rules and using one's discretion is difficult in every walk of life. A midrash (a commentary in the form of a story) on Genesis, written about the 5th century AD, says:

'Abraham said to God: "If you wish to maintain the world, strict justice is impossible; and if You want strict justice, the World cannot be maintained. You cannot hold the cord at both ends. You desire the world. You desire justice. Take one or the other. Unless you compromise the World cannot endure.".'

3.9 Cases when training is not the best answer

3.9.1 Electric plugs

Ordinary domestic three-pin electric plugs are sometimes wired incorrectly. This may be due to a slip, or to ignorance of the correct method.

The usual remedy is training and instruction: electric appliances come with instructions for fitting the plug. However, it is not difficult to reduce the opportunities for error. One method, now compulsory in the UK, is to fit the plug in the factory. Another method, so simple it is surprising that it has not been used more often, is to colour the terminals in the plug the same colour as the wires.

3.9.2 Kinetic handling

Training in kinetic handling methods is often recommended as a cure for back injuries but in many cases it is only part of the story. Before introducing a training programme the layout and design of areas where people have to work should be examined to check that they are the best possible. For example, a man has been seen working with his feet at different levels, lifting 30 lb from a conveyor belt onto staging behind him. A hoist was needed, not training of the man.

In another example there were two almost identical conveyor layouts, yet back accidents occurred on one and never on the other. An examination showed the first line was close to very large doors so that, when men stopped for a rest after getting hot and sweaty, they were standing in a draught, with the not unnatural result that they suffered back trouble.

In another example, trucks brought loads from stores and deposited them in unoccupied areas near to but not at the actual place where the goods would be required, with the result that men were then called upon to manhandle them over the last part. Very often the temporary resting places of these goods were pebbled areas and other unsatisfactory places so that men did not have a proper footing, with inevitable falls and strains as a result.

In short, we should examine the layout and planning of a task before considering training.

3.9.3 Attitudes

It is sometimes suggested that through training we should try to change people's attitudes. I doubt if such training is either justified or effective.

It is not justified because a person's attitude is his private affair. We should concern ourselves only with whether or not he achieves his objectives.

It is not effective because it is based on the assumption that if we change a person's attitude we change his behaviour. In fact, it is the other way round. An attitude has no real existence; it is just a generalization about behaviour.

If someone has too many accidents, let us discuss the reasons for them and the action needed to prevent them happening again. Behavioural safety training (see Section 5.3, page 107) can be very effective. After a while he may act differently, he may have fewer accidents and everyone will say that he has changed his attitude.

In short:

Don't try to change people's attitudes. Just help them with their problems, for example by tactfully pointing out unsafe acts and conditions.

William Blake wrote, 'He who would do good to another must do it in Minute Particulars. General Good is the plea of the scoundrel, hypocrite and flatterer.'[19] (See Section 5.2, page 103.)

Similarly, do not try to persuade Boards of Directors to change their policies. It is better to suggest ways of dealing with specific problems. Looking back, a common pattern may be seen. This is the policy — the common law of the organization.

References in Chapter 3

1. *Report of the President's Commission on the Accident at Three Mile Island* (the Kemeny Report), 1979 (Pergamon Press, New York, USA).
2. Kletz, T.A., 2001, *Learning from Accidents*, 3rd edition, Chapters 11 and 12 (Butterworth-Heinemann, Oxford, UK).
3. Marshall, E.E. *et al.*, 1981, *The Chemical Engineer*, No. 365, page 66.
4. *Petroleum Review*, July 1983, page 27.
5. *Petroleum Review*, July 1982, page 21.
6. *The Engineer*, 25 March 1966, page 475.
7. *Paris Match*, No. 875, 15 January 1966.
8. *Fire*, Special Supplement, February 1966.
9. *Petroleum Times*, 21 January 1966, page 132.
10. Lagadec, P., 1980, *Major Technological Risks*, page 175 (Pergamon Press, New York, USA).
11. Safety training packages, various dates (Institution of Chemical Engineers, Rugby, UK).
12. Kletz, T.A., 2001, *Learning from Accidents*, 3rd edition, Chapter 16 (Butterworth-Heinemann, Oxford, UK).
13. *Report of Royal Commission into the Failure of the West Gate Bridge*, 1971 (State of Victoria Government Printer, Melbourne, Australia).
14. Report issued by the US National Transportation Safety Board, Washington, DC, 1975.
15. *Petroleum Review*, January 1974.
16. Howarth, A., 1984, *Chemistry in Britain*, 20(2): 140.
17. *The Guardian*, 23 June 1971.
18. Sandvig, H.J., 1983, *JAOCS*, 60(2): 243.
19. Blake, W., *Jerusalem*, 55, 60.
20. Kletz, T.A., 1996, *Dispelling Chemical Engineering Myths*, 3rd edition (Taylor & Francis, Philadelphia, PA, USA).
21. *New York Times*, 19 November 1987.
22. Parker, R.J. (Chairman), 1975, *The Flixborough Disaster: Report of the Court of Inquiry* (HMSO, London, UK).

23. Kletz, T.A., 2001, *Learning from Accidents*, 3rd edition, Chapter 8 (Butterworth-Heinemann, Oxford, UK).
24. Tuchman, B.W., 1988, *The First Salute*, pages 120–128, 165 and 175 (Knopf, New York, USA).
25. Mackenzie, S.P., 1996, *The Home Guard*, page 74 (Oxford University Press, Oxford, UK).
26. *Operating Experience Weekly Summary*, 1999, No. 99–08, page 16 (Office of Nuclear and Facility Safety, US Department of Energy, Washington, DC, USA).
27. *Operating Experience Weekly Summary*, 1999, No. 99–04, page 1 (Office of Nuclear and Facility Safety, US Department of Energy, Washington, DC, USA).
28. Pratt, T.H. and Atherton, J.G., 1999, *Process Safety Progress*, 18(4): 241.
29. *Operating Experience Weekly Summary*, 1999, No. 99–15, page 2 (Office of Nuclear and Facility Safety, US Department of Energy, Washington, DC, USA).
30. Ward, R., 1995, *Chemical Technology Europe*, 2(4): 26.
31. *Operating Experience Weekly Summary*, 1998, No. 98–27, page 5 (Office of Nuclear and Facility Safety, US Department of Energy, Washington, DC, USA).
32. *Operating Experience Weekly Summary*, 1998/1999, No. 98–15, page 3 and No. 99–16, page 5 (Office of Nuclear and Facility Safety, US Department of Energy, Washington, DC, USA).
33. Bryson, B., 1998, *Notes from a Big Country*, page 104 (Doubleday, London, UK).
34. McClure, A., 1998, *Daily Telegraph*, date unknown.
35. Lindley, J., August 1997, *Loss Prevention Bulletin*, No. 136, page 7.
36. *Inquiry into the Fire on Heavy Goods Vehicle Shuttle 7539 on 18 November 1996*, May 1997 (HMSO, London, UK).
37. Williams, T. and Gromacki, M., 1999, Eliminating error-likely situations during procedure updates, *American Institute of Chemical Engineers 33rd Annual Loss Prevention Symposium, 15–17 March 1999.*
38. *Operating Experience Weekly Summary*, 1999, No. 99–34, page 1 (Office of Nuclear and Facility Safety, US Department of Energy, Washington, DC, USA).
39. Stephens, D., *Natural History*, July/August 2000, page 38.

Accidents due to a lack of physical or mental ability

4

'Nothing is impossible for people who do not have to do it themselves.'
Anon

These accidents are much less common than those described in Chapters 2 and 3 but nevertheless do occur from time to time. Most of them occur because people are asked to do more than people as a whole are capable of doing, physically or mentally. Only a few occur because someone was asked to do more than he (or she) was individually capable of doing.

To prevent the accidents described in this chapter all we can do is to change the work situation — that is, the design or method of working.

4.1 People asked to do the physically difficult or impossible

(1) A steel company found that overhead travelling magnet cranes were frequently damaging railway wagons. One of the causes was found to be the design of the crane cab. The driver had to lean over the side to see his load. He could then not reach one of the controllers, so he could not operate this control and watch the load at the same time[1].

(2) A refinery compressor was isolated for repair and swept out with nitrogen but, because some hydrogen sulphide might still be present, the fitters were told to wear air-line breathing apparatus. They found it difficult to remove a cylinder valve which was situated close to the floor, so one fitter decided to remove his mask and was overcome by the hydrogen sulphide. Following the incident lifting aids were provided.

Many companies would have been content to reprimand the fitter for breaking the rules. The company concerned, however, asked why he had removed his mask and it then became clear that he had been asked to carry out a task which was difficult to perform while wearing a mask[2].

(3) Incidents have occurred because valves which have to be operated in an emergency were found to be too stiff. Such valves should be kept lubricated and exercised from time to time.

(4) Operators often complain that valves are inaccessible. Emergency valves should, of course, always be readily accessible. But other valves, if they have to be operated, say, once per year or less often, can be out of reach. It is reasonable to expect operators to fetch a ladder or scramble into a pipe trench on rare occasions.

Designers should remember that if a valve is just within reach of an average person then half the population cannot reach it. They should design so that 95% (say) of the population can reach it.

(5) It is difficult to give lorry drivers a good view while they are reversing. Aids such as large mirrors should be provided in places such as loading bays where a lot of reversing has to be done[1].

(6) Related to these accidents are those caused by so-called clumsiness. For example, an electrician was using a clip-on ammeter inside a live motor-starter cubicle. A clip-on ammeter has two spring-loaded jaws which are clipped round a conductor forming a coil which measures the current by induction. The jaws are insulated except for the extreme ends. The electrician accidentally shorted two live phases (or a live phase and earth) with the bare metal ends of the jaws. He was burnt on the face and neck and the starter was damaged[3]. In many companies such an accident would be put down to 'clumsiness' and the man told to take more care.

Such accidents are the physical equivalent of the mental slips discussed in Chapter 2. The method of working makes an occasional accident almost inevitable. Sooner or later, for one reason or another, the electrician's co-ordination of his movements will be a little poorer than normal and an accident will result. We all have off days. The method of working is hazardous — though accepted practice — and a better method should be sought.

When handling hazardous materials or equipment we usually provide at least two safeguards (defence in depth). In this case the only safeguard was the electrician's skill. Another, physical safeguard should be provided[19].

(7) The following is another example of what some people might call clumsiness but was in fact poor layout. A man was asked to vent gas at a gauge pressure of 16 bar (230 psi). The vent valve, a ball valve, was stiff and had no handle and so he used a wrench with a pipe extension on the handle. While opening the valve he accidentally moved in front of the opening and was thrown 2 m (6 feet)

by the escaping gas. He would have been thrown further if a tank had not been in the way. The vent valve should have been fitted with a tail pipe to direct the escaping gas in a safe direction[26].

The missing tail pipe and handle could have been spotted during plant inspections or anyone who saw them could have reported them. Whether or not they do so not depends on the encouragement they receive and on whether or not such reports are acted on.

(8) A leak of petrol from an underground tank at a service station went undetected for two weeks by which time 10,000 gallons (50 m^3) had leaked into the surrounding ground and contaminated a stream and a river. When the staff were asked why they did not carry out a daily check by dipping, as they had been told to do, it was found that the manhole cover over the dip pipe was heavy and difficult to move. It was replaced by one with a small central disc that could be removed easily[20].

(9) When train drivers pass a signal at red it is usually due to a lapse of attention (Section 2.9.3, page 38). One case, however, was due to diabetic eye disease that had reduced the driver's colour vision. He was aware of his condition but had not reported it[27].

4.2 People asked to do the mentally difficult or impossible

Many of the incidents described in Chapter 2 almost come into this category. If a man is told to close a valve when an alarm sounds or at a particular stage in an operation he cannot be expected to close the right valve in the required time *every time*. His error rate will be higher if the layout or labelling are poor or he is under stress or distracted. However, he will close the right valve most of the time. This section considers some accidents which occurred because people were expected to carry out tasks which most people would fail to carry out correctly on many occasions. These failures are of several sorts: those due to information or task overload or underload; those involving detection of rare events; those involving habits; and those involving estimates of dimensions.

4.2.1 Information or task overload

A new, highly-automated plant developed an unforeseen fault. The computer started to print out a long list of alarms. The operator did not know what had occurred and took no action. Ultimately an explosion occurred.

Afterwards the designers agreed that the situation should not have occurred and that it was difficult or impossible for the operator to diagnose the fault, but they then said to him, 'Why didn't you assume the worst and trip the plant? Why didn't you say to yourself, "I don't know what's happening so I will assume it is a condition that justifies an emergency shut-down. It can't be worse than that"?'

Unfortunately people do not think like that. If someone is overloaded by too much information he may simply switch off (himself, not the equipment) and do nothing. The action suggested by the designers may be logical, but this is not how people behave under pressure (see also Section 12.4, page 210). Reference 37 describes methods to reduce alarm overload.

The introduction of computers has made it much easier than in the past to overload people with too much information, in management as well as operating jobs. If quantities of computer print-out are dumped on people's desks every week, then most of it will be ignored, including the bits that should be looked at.

Plant supervisors sometimes suffer from task overload — that is, they are expected to handle more jobs at once than a person can reasonably cope with. This has caused several accidents. For example, two jobs had to be carried out simultaneously in the same pipe trench, 20 m apart. The first job was construction of a new pipeline. A permit-to-work, including a welding permit, was issued at 08.00 hours, valid for the whole day. At 12.00 hours a permit was requested for removal of a slip-plate from an oil line. The foreman gave permission, judging that the welders would by this time be more than 15 m (50 ft) from the site of the slip-plate. He did not visit the pipe trench which was 500 m away, as he was dealing with problems on the operating plant. Had he visited the trench he might have noticed that it was flooded. Although the pipeline had been emptied, a few gallons of light oil remained and ran out when the slip-plate joint was broken. It spread over the surface of the water in the pipe trench and was ignited by the welders. The man removing the slip-plate was killed.

The actual distance between the two jobs — 20 m — was rather close to the minimum distance — 15 m — normally required. However the 15 m includes a safety margin. Vapour from a small spillage will not normally spread anything like this distance. On the surface of water, however, liquid will spread hundreds of metres.

Afterwards a special day supervisor was appointed to supervise the construction operations. It was realized that it was unrealistic to expect the foreman, with his primary commitment to the operating plant, to give the construction work the attention it required.

If extensive maintenance or construction work is to be carried out on an operating plant — for example, if part of it is shut down and the rest running — extra supervision should be provided. One supervisor should look after normal plant operations and the other should deal with the maintenance or construction organization.

Overloading of a supervisor at a busy time of day may have contributed to another serious accident. A shift process supervisor returned to work after his days off at 08.00 hours on Monday morning. One of his first jobs was to issue a permit-to-work for repair of a large pump. When the pump was dismantled, hot oil came out and caught fire. Three men were killed and the plant was destroyed. It was found that the suction valve on the pump had been left open.

The supervisor said he had checked the pump before issuing the permit and found the suction valve (and delivery valve) already shut. It is possible that it was and that someone later opened it. It is also possible that the supervisor, due to pressure of work, forgot to check the valve.

The real fault here, of course, is not the overloading of the supervisor — inevitable from time to time — but the lack of a proper system of work. The valves on the pump should have been locked shut and in addition the first job, when maintenance started, should have been to insert slip-plates. If valves have to be locked shut, then supervisors have to visit the scene in order to lock them.

Overloading of shift supervisors can be reduced by simple arrangements. They should not, for example, start work at the same time as the maintenance team.

4.2.2 Detection of rare events

If a man is asked to detect a rare event he may fail to notice when it occurs, or may not believe that it is genuine. The danger is greatest when he has little else to do. It is very difficult for night-watchmen, for example, to remain alert when nothing has been known to happen and when there is nothing to occupy their minds and keep them alert. Compare Jesus' remark to the disciples who had fallen asleep, 'Could ye not watch with me one hour?' (St. Matthew, 26, 40). On some of the occasions when train drivers have passed a signal at danger (see Section 2.9.3, page 38), the drivers, in many years of driving regularly along the same route, had never before known that signal to be at danger.

Similarly, in the days before continuously braked trains became the rule, railway signalmen were expected to confirm that each train that passed the signal box was complete, as shown by the presence of tail lights.

In 1936 an accident occurred because a signalman failed to spot the fact that there were no tail lights on the train. In his 25 years as a signalman he had never previously had an incomplete train pass his box[4]. The signalman was blamed (in part) for the accident but it is difficult to believe that this was justified. It is difficult for anyone to realize that an event that has never occurred in 25 years has in fact just happened.

In September 1984 the press reported that after a large number of false alarms had been received, an ambulance crew failed to respond to a call that two boys were trapped in a tank. The call was genuine (see also Section 5.2.7, page 106).

4.2.3 Task underload

Reliability falls off when people have too little to do as well as when they have too much to do. As already stated, it is difficult for night-watchmen to remain alert.

During the Second World War, studies were made of the performance of watchkeepers detecting submarines approaching ships. It was found that the effectiveness of a man carrying out such a passive task fell off very rapidly after about 30 minutes.

It is sometimes suggested that we should restrict the amount of automation on a plant in order to give the operators enough to do to keep them alert. I do not think this is the right philosophy. If automation is needed to give the necessary reliability, then we should not sacrifice reliability in order to find work for the operators. We should look for other ways of keeping them alert. Similarly if automation is chosen because it is more efficient or effective, we should not sacrifice efficiency or effectiveness in order to find work for the operators.

In practice, I doubt if process plant operators often suffer from task underload to an extent that affects their performance. Although in theory they have little to do on a highly automated plant, in practice there are often some instruments on manual control, there are non-automated tasks to be done, such as changing over pumps and tanks, there is equipment to be prepared for maintenance, routine inspections to be carried out, and so on.

If, however, it is felt that the operators are seriously underloaded, then we should not ask them to do what a machine can do better but look for useful but not essential tasks that will keep them alert and which can be set aside if there is trouble on the plant — the process equivalent of leaving the ironing for the babysitter.

One such task is the calculation and graphing of process parameters such as efficiency, fuel consumption, catalyst life and so on.

Another task is studying a training programme on a video or CD-ROM (but not on the control computer). This is best done under the supervision of a foreman who stops the programme at suitable intervals, adds extra explanation if necessary and discusses it with the operators.

Despite what has been said above, fully operated plants are not necessarily the most reliable. Hunns[5] has compared three designs for a boiler control system: a largely manual design, a partly-automated design and a fully automated design. The comparison showed that the partly-automatic design was the most reliable. Section 7.5.2 (page 143) shows that an operator may be more reliable at some process tasks than automatic equipment.

More important than occupying the operator's time, is letting him feel that he is in charge of the plant — able to monitor it and able to intervene when he considers it necessary — and not just a passive bystander watching a fully automatic system. The control system should be designed with this philosophy in mind[6] (see also Section 12.4, page 210).

In deciding whether to give a task to a person or a machine we should give it to whichever is least likely to fail. Machines are better than people at boring tasks such as monitoring equipment, repetitive tasks, tasks which consist of many simple elements (as people have many opportunities for slips or lapses of attention) and tasks which require very rapid response. People are necessary for dealing with unforeseen situations, situations which are not fully defined (that is, all the necessary information is not available) and situations where the action cannot be specified in advance[28].

4.2.4 Habits and expectations

If we expect people to go against established habits, errors will result. For example, if a man is trained to drive a vehicle with the controls laid out in one way and is then transferred to another vehicle where the layout is different, he will probably make errors. A good example occurred in a company which was developing an attachment to fit on the back of a tractor. The driving controls had to be placed behind the seat so that the driver could face the rear. On the development model the control rods were merely extended to the rear and thus the driver found that they were situated in the opposite positions to those on a normal tractor. When someone demonstrated the modified tractor, he twice drove it into another vehicle[1].

Habits are also formed by general experience as well as experience on particular equipment. In the UK people expect that pressing a switch down will switch on a light or appliance. If a volume control is turned clockwise we expect the sound to get louder. If the controller on a crane is moved to the right we expect the load to move to the right. Errors will occur if designers expect people to break these habits.

Trials have shown that most people leaving a building in an emergency ignore emergency exit signs and instead leave by the routes they know best[29]. (If the normal route includes a double door, one half of which is kept closed, very few people unbolt it.)

A biologist studying the behaviour of migrating chirus (a relative of antelopes, native to China) noticed that they detoured round an area of grass. Ten thousand years ago the grassland was a lake and the chirus still seemed to follow their ancient, circuitous route[30].

Expectations are closely related to habits. A plant was fitted with a manually operated firewater sprinkler system. The valves were above the ceiling and were operated by chains. Operators checked that the valves were open by closing them slightly and then opening them fully. They reported that they were open when they were in fact closed. They were so sure that they would be unable to open them they did not try very hard. In a better system the valves would be visible and it would be possible to see at a glance whether they were open or closed[31].

4.2.5 Estimates of dimensions

People are not very good at estimating distances. Vehicles are often driven through openings which are too narrow or too short. Drivers should know the dimensions of their vehicles and the dimensions of narrow openings should be clearly displayed[1].

4.3 Individual traits and accident proneness

A few accidents occur because individuals are asked to do more than they are capable of doing, though the task would be within the capacity of most people.

For example, at a weekly safety meeting a maintenance foreman reported that one of his men had fallen off a bicycle. He had been sent to the store for an item that was required urgently and on the way, while entering a major road, had collided with another cyclist. The bicycle and road surface were in good condition, and the accident seemed to be a typical 'human failing' one.

Further inquiries, however, disclosed that the man was elderly, with poor eyesight, and unused to riding a bicycle. He had last ridden one when he was a boy. The foreman had sent him because no-one else was available.

In 1935 a signalman on a busy line accepted a second train before the first one had left the section. It came out in the inquiry that although he had 23 years' experience and had worked in four different boxes, he had been chosen for this present box by seniority rather than merit and had taken five weeks to learn how to work it[4].

A train driver passed a signal at danger and caused an accident in which one person was killed and 70 injured. The inquiry disclosed that between four and eight years earlier he had committed six irregularities. Three involved braking, one excessive speed and two passing signals at danger. His suitability as a driver was obviously suspect[32]. Most signal-passed-at-danger incidents, however, occur to normally good drivers who have a momentary lapse of attention (see Section 2.9.3, page 38).

This brings us to the question of accident proneness which has been touched on already in Section 2.1 (page 11). It is tempting to think that we can reduce accidents by psychological testing but such tests are not easy and few accidents seem to occur because individuals are accident-prone. Hunter[7] writes:

'Accident proneness is greatly influenced by the mental attitude of the subjects. The accident-prone are apt to be insubordinate, temperamentally excitable and to show a tendency to get flustered in an emergency. These and other defects of personality indicate a lack of aptitude on the part of the subjects for their occupation. But whenever we are considering how to prevent accidents we must avoid the danger of laying too much blame on abnormalities of temperament and personality. Let us beware lest the concept of the accident-prone person be stretched beyond the limits within which it can be a fruitful idea. We should indeed be guilty of a grave error if for any reason we discouraged the manufacture of safe machinery.'

Swain[8] describes several attempts to measure the contribution of accident-prone individuals to the accident rate. In one study of 104 railway shunters over three years, the ten shunters with the highest accident rate in the first year were removed from the data for the following two years. The accident rate for these two years actually rose slightly. Similar results were obtained in a study of 847 car drivers.

Eysenck[9], however, shows that personality testing of South African bus-drivers reduced accident rates by 60% over 10 years.

Whatever may be the case in some industries, in the process industries very few individuals seem to have more than their fair share of accidents. If anyone does, we should obviously look at his work situation and if it is normal, consider whether he is suitable for the particular job. However, remember that a worker may have more than the average share of accidents by chance. Suppose that in a factory of 675 workers there are in a given period 370 accidents. Obviously there are not enough accidents to go round and many workers will have no accidents and most of the others will have only one. If the accidents occur at random, then:

- 11 workers will have 3 accidents each;
- 1.5 workers will have 4 accidents each;
- 0.15 workers will have 5 accidents each.

That is, once in every six or seven periods we should expect one worker to have five accidents by chance.

Swain[8] describes a study of 2300 US railway employees. They were divided into 1828 low-accident men, who had four or less accidents, and 472 high-accident men, who had five or more accidents. If the accidents were distributed at random then we would expect 476 men to have five or more accidents.

In examining accident statistics it is, of course, necessary to define the boundaries of the sample before examining the figures. If we select a group of accidents and then fix the accident type, job definition and time period so that the number of accidents is unusually high, we can prove anything.

This is similar to the Texas sharpshooter who empties his gun at the barn door and then draws a target round the bullet holes to prove what a good shot he is.

Another reason why some men may have an unusually large number of accidents is that they deliberately but unconsciously injure themselves in order to have an excuse for withdrawing from a work situation which they find intolerable[10] (or, having accidentally injured themselves, use this as an excuse to withdraw from work). High accident rates in a company may even be a form of mass psychogenic illness, a mass reaction to poor working conditions[21].

Obviously changing the job so as to remove opportunities for accidents will not prevent accidents of this type. The people will easily find other ways of injuring themselves. If we believe that a man's injuries occur in this way then we have to try to find out why he finds work intolerable. Perhaps he does not get on with his fellow workers; perhaps his job does not provide opportunities for growth, achievement, responsibility and recognition (see Appendix 1).

To sum up on accident proneness, if a man has an unusually large number of accidents, compared with his fellows, this may be due to:

- chance;
- lack of physical or mental ability;
- personality;

- possibly, psychological problems. Accidents may be a symptom of with-drawal from the work situation.

The last three categories do not seem to contribute a great deal to accidents in the process industries.

In some countries accidents may occur because employees' command of the written or spoken language used is poor or because their education and background make them reluctant to ask questions; the culture of the home and the factory may not be the same[11].

4.4 Mind-sets

We have a problem. We think of a solution. We are then so busy congratulating ourselves that we fail to see that there may be a better solution, that some evidence points the other way, or that our solution has unwanted side-effects. This is known as a 'mind-set' or, if you prefer a more technical term, *Einstellung*.

'When a scenario fits the current situation, there is a tendency to believe that this scenario is the only one that can occur. Other scenarios may well be possible, but these are rejected automatically.'[22]

De Bono writes, 'Even for scientists there comes a point in the gathering of evidence when conviction takes over and thereafter selects the evidence.'[12]

Mind-sets are described by Raudsepp[13]:

'Most people when faced with a problem, tend to grab the first solution that occurs to them and rest content with it. Rare, indeed, is the individual who keeps trying to find other solutions to his problem. This is especially evident when a person feels under pressure ...'

'And once a judgement is arrived at, we tend to persevere in it even when the evidence is overwhelming that we are wrong. Once an explanation is articulated, it is difficult to revise or drop it in the face of contradictory evidence ...'

'Many interesting psychological experiments have demonstrated the fixating power of premature judgements. In one experiment, color slides of familiar objects, such as a fire hydrant, were projected upon a screen. People were asked to try to identify the objects while they were still out of focus. Gradually the focus was improved through several stages. The striking finding was this: If an individual wrongly identified an object while it was far out of focus, he frequently still could not identify it correctly when it was brought sufficiently into focus so that another person who had not seen the blurred vision could easily identify it. What this indicates is that considerably more effort and evidence is necessary to overcome an incorrect judgement, hypothesis or belief than it is to establish a correct one. A person who is in the habit of jumping to conclusions frequently closes his mind to new information, and limited awareness hampers creative solutions.'

Mind-sets might have been discussed under 'training', as the only way of avoiding them seems to be to make people aware of their existence and to discourage people from coming to premature conclusions. However, they are included in this chapter because they are a feature of people's mental abilities which is difficult to overcome and which we have to live with.

Here are some examples of mind-sets. Others are described in Sections 2.9, 3.2 and 10.2 (pages 35, 50 and 178).

4.4.1 An operator's mind-set

A good example is provided by an accident that occurred many years ago, on the coker shown, in a simplified form, in Figure 4.1[14].

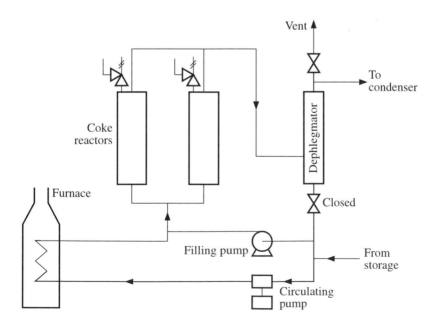

Figure 4.1 Simplified line diagram of coker

The accident occurred while the unit was being started up following a shutdown to empty the coker of product, an operation that took place every few days. The normal procedure was to fill the plant with oil by opening the vent on the top of the dephlegmator and operating the low pressure filling pump and high pressure circulating pump in parallel. When oil came out of the vent, it was closed, the filling pump shut down, circulation established and the furnace lit. This procedure, though primitive, was not unsafe as the oil had a high flash-point (32°C).

On the night of the accident the afternoon shift operator forgot to open the vent. When the night shift came on duty at 11.00 pm they found that the plant was filling with oil more slowly than usual and that the pump delivery pressure was higher than usual. As it was a very cold night they decided (about 2.00 am) that the low pumping rate and high pressure were due to an unusually high viscosity of the oil and they decided to light one burner in the furnace. Their diagnosis was not absurd, merely wrong. On earlier occasions lighting a burner had cured the same symptoms.

On this occasion it did not. The operators, however, were reluctant to consider that their theory might be wrong.

The filling pump got hot and had to be shut down. It was pumping against a rising pressure, which it could not overcome. The operators, however, ignored this clue and blamed the overheating of the pump on poor maintenance.

Finally, at about 5.00 am the pump delivery pressure started to rise more rapidly. (Though far above normal for the filling stage, it was below the normal on-line operating pressure and so the relief valves did not lift.) The operators at last realized that their theory might be wrong. They decided to check that the vent valve was open. They found it shut. Before they could open it an explosion occurred killing one of the operators.

The cause of the explosion is interesting, though not related to the subject of this book. The dephlegmator acted as a giant slow-moving diesel engine, the rising level of oil compressing the air and oil vapour above it and raising their temperature until it exceeded the auto-ignition temperature of the oil vapour.

The incident is also interesting as an example of the slips discussed in Chapter 2. On the plant and a neighbouring one there had been 6000 successful start-ups before the explosion. The design and method of working made an error in the end almost inevitable but the error rate (1 in 6000) was very low, lower than anyone could reasonably expect. (1 in 1000 would be a typical figure.) However it may be that the vent valve had been left shut before but found to be so in time, and the incident not reported. If it had been reported, other shifts might have been more aware of the possible error.

4.4.2 A designer's mind-set

A new tank was designed for the storage of refrigerated liquid ethylene at low pressure (a gauge pressure of 0.8 psi or 0.05 bar). Heat leaking into the tank would vaporize some ethylene which was to be cooled and returned to the tank. When the refrigeration unit was shut down — several days per year — the ethylene in the tank would be allowed to warm up a few degrees so that the relief valve lifted (at a gauge pressure of 1.5 psi or 0.1 bar) and the vapour would be discharged to a low stack (20 m high).

After construction had started, it was realized that on a still day the cold, heavy ethylene vapour would fall to ground level where it might be ignited. Various solutions were considered and turned down. The stack could not be turned into a flare stack, as the heat radiation at ground level would be too high. The stack could not be made taller as the base was too weak to carry the extra weight. Finally, someone had an idea: put steam up the stack to warm up the ethylene so that it continued to rise and would not fall to the ground. This solution got everyone off the hook and was adopted.

When the plant was commissioned, condensate from the steam, running down the walls of the stack, met the cold ethylene gas and froze, completely blocking the 8 inch diameter stack. The tank was overpressured and split near the base. Fortunately the escaping ethylene did not ignite and the tank was emptied for repair without mishap. Afterwards a flare stack was constructed.

In retrospect, it seems obvious that ethylene vapour at −100°C might cause water to freeze, but the design team were so hypnotized by their solution that they were blind to its deficiencies. Use of a formal technique, such as a hazard and operability study[15], for considering the consequences of changes would probably have helped.

This incident is interesting in another respect. For 11 hours before the split occurred the operators (on two shifts) were writing down on the record sheet readings above the relief valve set point; they steadied out at a gauge pressure of 2 psi (0.14 bar), the full-scale deflection of the instrument. They did not realize the significance of the readings, and took no action. They did not even draw the foremen's attention to them and the foremen did not notice the high readings on their routine tours of the plant.

Obviously better training of the operators was needed, but in addition a simple change in the work situation would help: readings above which action is required should be printed, preferably in red, on the record sheets.

4.4.3 A classical mind-set

If marble columns lie on the ground while awaiting erection, they may become discoloured. In Renaissance times, therefore, they were often supported on two wooden beams as shown in Figure 4.2(a). This had the additional advantage that it was easier to get a lifting rope underneath them. Someone realized that a long column might break under its own weight, as shown in Figure 4.2(b). A third beam was therefore added (Figure 4.2(c)). No-one realized that if one of the end beams sank into the ground then the cantilevered end might break under its own weight, as shown in Figure 4.2(d).

Petroski comments that, as with so many design changes, the modification was made to overcome one problem but created another problem. It is the designer's responsibility, he says, to see how a change interacts with the rest of the existing system[33].

4.4.4 A simple mind-set

Riggers were trying to lift, with a crane, a casting weighing 180 kg (400 lb). They had been told that it had been unbolted from its support but it did not budge. Nevertheless, they continued to try to raise it. On the third attempt the

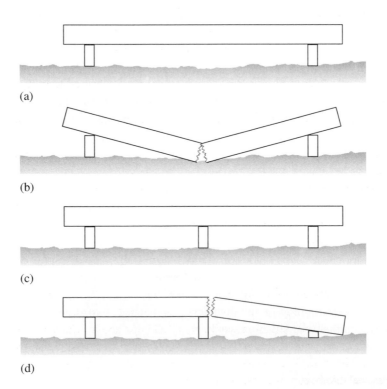

(a)

(b)

(c)

(d)

Figure 4.2 Two ways of supporting a marble column so that it is not in contact with the ground, and possible failure modes

nylon sling broke, though it was rated for 5 tonnes. This caused the crane hook to swing, fortunately without causing any injury or damage. The casting was in fact still secured and in addition had sharp edges that may have cut the sling. There was no load cell to indicate the force being applied by the crane[34].

4.4.5 Scholarly mind-sets

Mind-sets are not restricted to operators and designers. There are many examples of distinguished scholars who, having adopted a theory, continue to believe in it even though the evidence against it appears, to others, to be overwhelming. Here are two examples. Some camel-skin parchments were offered for sale in Jordan in 1966. The writing was similar to early Hebrew but not identical with it. A distinguished biblical scholar dated them to the 9th–7th centuries BC — 600 or 700 years earlier than the Dead Sea Scrolls — and suggested that the language was Philistine. If so, they would be the first documents yet discovered in that language.

Later, other scholars pointed out that the writing was a copy of a well-known early Hebrew inscription, but with the words in the wrong order. The documents were obviously fakes. The original identifier of the documents stuck to his opinion.

The parchments were dated by radio-carbon tests. These showed that they were modern. The scholar then said that the tests were meaningless; the parchments had been handled by so many people that they could have been contaminated[16].

One classical scholar, convinced that the ancient Greeks could write nothing but great poetry, thought he had discovered the poetic metrical system for the poetry of the Linear B tablets — no mean achievement when one realizes that these tablets are administrative texts connected with the wool industry, the flax industry, coppersmithing and the manufacture of perfumes and unguents[17].

Miller writes[35]:

'... once an idea lodges in the imagination, it can successfully eliminate or discredit any evidence which might be regarded as contradictory ...

'The history of science is full of such examples. If a theory is persuasive enough ... scientists will accommodate inconsistent or anomalous findings by decorating the accepted theory with hastily improvised modifications. Even when the theory has become an intellectual slum, perilously propped and patched, the community will not abandon the condemned premises until alternative accommodation has been developed.'

4.4.6 Cognitive dissonance

These examples illustrate what is sometimes called 'cognitive dissonance' — literally, an unpleasant noise in the mind[23]. If new information upsets our beliefs or means that we have to change our ways, we deny it or minimize its importance. What should we do when we meet this reaction?

If time permits, we should be patient. We cannot expect people to discard in a few minutes, or even a few weeks, the beliefs they have held for years.

If we believe that incorrect diagnosis of a plant problem may cause an accident, or operating difficulties, there may be no time for patience. We have to persuade those concerned that another scenario is possible. The diagnosis is probably recent and will not be as deeply ingrained as views that have been held for a long time.

The historian Barbara Tuchman writes that in wartime, 'It is very difficult for a recipient of secret information to believe its validity when it does not conform to his preconceived plans or ideas; he believes what he wants to believe and rejects what does not verify what he already knows, or thinks he knows,' and, 'preconceived notions can be more damaging than cannon.'[24]

4.4.7 Notorious mind-set

A frightening example of people's ability to see only the evidence that supports their view and ignore the rest is provided by the story of the witches of Salem, as told by Marion Starkey[18]. In the 18th century several people in the town of Salem in New England were accused of witchcraft and executed and for a time hysteria gripped the town. Even after the original accusers had suspiciously fled some of those in authority were unwilling to admit that they might have been wrong.

Similar incidents occurred in England, and more frequently in Scotland, at about the same time. Until the 13th century the Church considered the traditional power of witches as imaginary, bred of insanity or dreams; it was heretical to express belief in their reality[36]. We are not always more enlightened than our forebears.

As Paul Johnson writes (in a different context), 'As with all conspiracy theories, once the first imaginative jump is made, the rest follows with intoxicating logic.'[25]

References in Chapter 4

1. Sell, R.G., January 1964, *Ergonomics Versus Accidents* (Ministry of Technology, London, UK).
2. *Petroleum Review*, April 1982, page 34.

3. *Petroleum Review*, September 1984, page 33.
4. Gerard, M. and Hamilton, J.A.B., 1984, *Rails to Disaster*, pages 72 and 59 (Allen and Unwin, London, UK).
5. Hunns, D.M., 1981, *Terotechnica*, 2: 159.
6. Love, J., 1984, *The Chemical Engineer*, No. 403, page 18.
7. Hunter, D., 1975, *The Diseases of Occupations*, 5th edition, page 1064 (English Universities Press, Sevenoaks, UK).
8. Swain, A.D., A work situation approach to job safety, in Widner, J.T. (editor), 1973, *Selected Readings in Safety*, page 371 (Academy Press, Macon, Georgia, USA).
9. Eysenck, H.J., 1965, *Fact and Fiction in Psychology*, Chapter 6 (Penguin Books, London, UK).
10. Hill, J.M.M. and Trist, E.L., 1962, *Industrial Accidents, Sickness and Other Absences*, Pamphlet No 4 (Tavistock Publications, London, UK).
11. Foxcroft, A., July 1984, *Safety Management* (South Africa), page 29.
12. de Bono, F., 1981, *An Atlas of Management Thinking*, page 129 (Maurice Temple Smith, Aldershot, UK) (also published by Penguin Books, London, 1983).
13. Raudsepp, E., 1981, *Hydrocarbon Processing*, 60(9): 29.
14. Pipkin, O.A. in Vervalin, C.H. (editor), 1985, *Fire Protection Manual for Hydrocarbon Processing Plants*, 3rd edition, Volume 1, page 95 (Gulf Publishing Co, Houston, Texas, USA).
15. Kletz, T.A., 1999, *Hazop and Hazan — Identifying and Assessing Process Industry Hazards*, 4th edition (Institution of Chemical Engineers, Rugby, UK).
16. *Biblical Archaeology Review*, 1984, 10(3): 66.
17. Muhly, J.D., 1983, *Biblical Archaeology Review*, 9(5): 74.
18. Starkey, M.L., 1949, *The Devil in Massachusetts* (Knopf, New York, USA) (also published by Anchor Books, New York, USA, 1969). The witches of Salem are also the subject of Arthur Miller's play, *The Crucible*.
19. Bond, J. and Bryans, J.W., 1987, *Loss Prevention Bulletin*, No. 075, page 27.
20. *Safety Management* (South Africa), September 1989, page 47.
21. Schmitt, N. and Fitzgerald, M., 1982, *Psychogenic Illness* (Erlbaum, New Jersey, USA).
22. Henley, E.J. and Kumamoto, H., 1985, *Designing for Reliability and Safety Control*, page 335 (Prentice-Hall, Englewood Cliffs, New Jersey, USA).
23. Kletz. T.A., 1990, *Critical Aspects of Safety and Loss Prevention*, page 69 (Butterworths, London, UK).
24. Tuchman, B.W., 1988, *The First Salute*, page 254 (Knopf, New York, USA).
25. Johnson, P., 1987, *A History of the Jews*, page 211 (Weidenfeld and Nicholson, London, UK).
26. *Operating Experience Weekly Summary*, 1999, No. 99–20, page 7 (Office of Nuclear and Facility Safety, US Department of Energy, Washington, DC, USA).
27. *Lloyds Casualty Week,* 18 April 1977, quoted in *Disaster Prevention and Management,* 1997, 6(4): 267.

28. Study Group on the Safety of Operational Computer Systems, 1998, *The Use of Computers in Safety Critical Applications*, page 13 (HSE Books, Sudbury, UK).

29. Marsden, A.W., 1998, *Disaster Prevention and Management,* 7(5): 401.

30. Schaller, G.B., *Natural History,* May 1996, page 48.

31. *Operating Experience Weekly Summary*, 1998, No. 98–50, page 9 (Office of Nuclear and Facility Safety, US Department of Energy, Washington, DC, USA).

32. Graves, D., *Daily Telegraph,* 28 October 1996, page 5.

33. Petroski, H., 1994, *Design Paradigms*, Chapter 14 (Cambridge University Press, Cambridge, UK).

34. *Operating Experience Weekly Summary*, 1999, No. 99–21, page 7 (Office of Nuclear and Facility Safety, US Department of Energy, Washington, DC, USA).

35. Miller, J., 1978, *The Body in Question*, pages 189 and 190 (Cape, London, UK).

36. Peel, E. and Southern, P., 1969, *The Trials of the Lancashire Witches* (David & Charles, Newton Abbot, UK) (also published by Hendon Publishing, Nelson, UK, 1985).

37. *Alarm Systems, a Guide to Design, Management and Procurement*, 1999 (Engineering Equipment and Materials Users Association, London, UK).

Accidents due to failures to follow instructions

5

'When I don't do it, I am lazy; when my boss doesn't do it, he is too busy.'
Anon

This chapter considers some accidents which occurred not because of slips, ignorance or lack of ability, but because of a deliberate decision to do or not to do something. These decisions are often called violations but many are errors of judgement; sometimes people break the rules to make life easier for themselves but at other times they do so with the intention of benefiting the organization as a whole. Non-compliance is therefore a better term than violation.

These decisions are divided below into those made by managers and those made by operators. If operators cut corners it may be because they are not convinced that the procedure is necessary, in which case the accident is really due to a lack of training (see Chapter 2). They may however cut corners because all people carrying out a routine task become careless after a time. Managers should keep their eyes open to see that the proper procedures are being followed.

The division into two categories is not, of course, absolute. Junior managers may need training themselves, if they are not convinced of the need for safety procedures, and they may also cut corners if *their* bosses do not check up from time to time.

As with the slips and lapses of attention discussed in Chapter 2, whenever possible we should look for engineering solutions — designs which are not dependent on the operator carrying out a routine task correctly — but very often we have to adopt a software solution, that is, we have to persuade people to follow rules and good practice.

The errors discussed in this chapter and the next are the only ones in which people have any choice and in which blame has any relevance. Even here, before blaming people for so-called violations, we should ask:

- were the rules known and understood?
- was it possible to follow them?
- did they cover the problem?
- were the reasons for them known?

- were earlier failures to follow the rules overlooked?
- was he or she trying to help?
- if there had not been an accident, would he or she have been praised for their initiative?

There is a fine line between showing initiative and breaking the rules.

'The carefully considered risk, well thought out and responsibly taken, is commended ... Unless it fails, when like as not, it will be regarded as a damn stupid thing to have done in the first place. The taking of risks without proper responsible thought is reckless! ... Unless it comes off, when like as not, it will be regarded as inspired.'[12]

The English language contains a number of irregular verbs such as:

- I am firm;
- you are stubborn;
- he is pig-headed;

to which we can add:

- I show initiative;
- you break the rules;
- he is trying to wreck the job.

The examples that follow illustrate the ways in which non-compliance can be prevented or reduced. We should:

- Explain the reasons for the rules and procedures. We do not live in a society in which people will automatically do what they are told. They want to be convinced that the actions are necessary. The most effective way of convincing them is to describe and discuss accidents that would not have occurred if the rules had been followed. Rules imposed by authority rather than conviction soon lapse when the boss moves on or loses interest.
- Make sure everyone understands the rules. Do not send them through the post but explain them to and discuss them with the people who will have to carry them out.
- If possible, make the task easier. If the correct way of carrying out a job is difficult and there is an easier, incorrect way, the incorrect way will be followed as soon as our backs are turned. If people have to walk up and down stairs several times to operate a series of valves, there is a great temptation to operate them in the wrong order. We can tell them that this lowers the plant efficiency but moving the valves or operating them remotely will be more effective.

98

Another example: due to pressure of work a maintenance section were not carrying out the routine examination of flame arresters. The design was changed so that the job could be done by operators.

- Carry out checks and audits to make sure that the rules are being followed. We are not expected to stand over people all the time but we are expected to make occasional checks and we should not turn a blind eye when we see someone working unsafely (see Section 13.2, page 223).

To repeat the spelling example of Section 1.4 (page 6), I should be blamed for writing 'thru' only if there is a clear instruction not to use American spelling and no-one turned a blind eye when I used it in the past.

In a survey of the reasons for non-compliance[13], employees said that the following were the most important:

- if followed to the letter the job wouldn't get done;
- people are not aware that a procedure exists;
- people prefer to rely on their skills and experience;
- people assume they know what is in the procedure.

They recommended the following strategies for improvement:

- involving users in the design of procedures;
- writing procedures in plain English;
- updating procedures when plant and working practices change;
- ensuring that procedures always reflect current working practices (see Section 3.7, page 67).

5.1 Accidents due to non-compliance by managers

Chapter 6 considers the errors of senior managers who do not realize that they could do more to prevent accidents. This section is concerned with deliberate decisions not to carry out actions which someone has clearly been told to take or which are generally recognized as part of the job.

Accidents of this type have been particularly emphasized by W.B. Howard, one of the best-known US loss prevention engineers, particularly in a paper sub-titled 'We aint farmin' as good as we know how'[1]. He tells the story of a young graduate in agriculture who tried to tell an old farmer how to improve his methods. The farmer listened for a while and then said, 'Listen son, we aint farmin' now as good as we know how.'

Howard describes several accidents which occurred because managers were not 'farming' as well as they could — and should — have done. For example:

An explosion occurred in a hold tank in which reaction product had been kept for several hours. It was then found that:

- The pressure had been rising for several hours before the explosion but the operator on duty that day had had no training for the operation and did not realize the significance of the pressure rise.
- No tests had been made to see what happened when the reaction product was kept for many hours. Tests had been repeatedly postponed because everyone was 'too busy' running the unit.

The accident was not the result of equipment failure or human error as usually thought of — that is, a slip or lapse of attention (see Chapter 2) — but rather the result of conscious decisions to postpone testing and to put an inexperienced operator in charge.

Howard also described a dust explosion and fire which occurred because the operating management had bypassed all the trips and alarms in the plant, in order to increase production by 5%. Before you blame the operating management, remember they were not working in a vacuum. They acted as they did because they sensed that their bosses put output above safety. In these matters official statements of policy count for little. Little things count for more. When senior managers visited the plant, did they ask about safety or just about output? (See Section 3.5, page 65.)

If someone telephoned the head office to report an incident, what was the first question? Was it, 'Is anyone hurt?' or was it, 'When will you be back on line?'

Many years ago, when I was employed on operations, not safety, my works manager changed. After a few months, someone told the new works manager that the employees believed him to be less interested in safety than his predecessor. He was genuinely shocked. 'Whatever have I said,' he asked, 'to give that impression? Please assure everyone that I am fully committed to safety.' It was not what he had said that created the impression but what he had not said. Both managers spent a good deal of their time out on the site, and were good at talking to operators, supervisors and junior managers. The first works manager, unlike the second, frequently brought safety matters into the conversation.

Another example of a deliberate management decision was described in *Hansard*[2]. After a disgruntled employee had blown the whistle, a factory inspector visited a site where there were six tanks containing liquefied flammable gas. Each tank was fitted with a high level alarm and a high level trip. The factory inspector found that five of the alarms were not working, that no-one knew how long they had been out of order and that the trips were never tested, could not be tested and were of an unreliable design.

100

In this case the top management of the organization concerned was committed to safety, as they employed a large team of safety specialists in their head office, estimating accident probabilities. (Their calculations were useless as the assumptions on which they were based — that trips and alarms would be tested — were not correct. They would have been better employed out on the plant testing the alarms.) However the top management had failed to get the message across to the local management who took a deliberate decision not to test their trips and alarms.

When managers make a deliberate decision to stop an activity, they usually let it quietly lapse and do not draw attention to the change. However one report frankly stated, 'The data collection system … was run for a period of about 3 years … The system is now no longer running, not because it was unsuccessful, but because the management emphasis on the works has changed. The emphasis is now on the reduction of unnecessary work, not on the reduction of breakdowns.'[3]

Accidents are sometimes said to be system failures or organizational failures but systems and organizations have no minds of their own. Someone, a manager, has to change the system or organization. The accident report should say who should do so, in which way and by when, or nothing will happen.

5.1.1 Chernobyl

Chernobyl, a town in the Ukraine, was the scene of the world's worst nuclear accident, in 1986, when a water-cooled reactor overheated and radioactive material was discharged to the atmosphere. Although only about 30 people were killed immediately, several thousand more may die during the next 30 years, a one-millionth increase in the death rate from cancer in Europe. However, this forecast is based on the pessimistic assumption that the risk is proportional to the dose even when the dose is a small fraction of the normal background radiation.

There were two major errors in the design of the Chernobyl boiling water reactor, a design used only in the former Soviet Union:

(1) The reactor was unstable at outputs below 20%. Any rise in temperature increased the power output and the temperature rapidly rose further. In all other commercial nuclear reactor designs a rise in temperature causes a fall in heat output. The Chernobyl reactor was like a marble balanced on a convex surface. If it started to move, gravity made it move increasingly fast. Other reactors are like a marble on a concave surface (Figure 5.1, page 102). If the temperature rises the heat output falls.

Chernobyl   Other nuclear reactors

Figure 5.1 The Chernobyl nuclear reactor, at low outputs, was like a marble on a convex surface. If it started to move, gravity made it move increasingly fast. Other reactors are like a marble on a concave surface.

(2) The operators were told not to go below 20% output but there was nothing to prevent them doing so.

The accident happened during an experiment to see if the reactor developed enough power, while shutting down, to keep auxiliary equipment running during the minute or so that it took for the diesel-driven emergency generators to start up. There was nothing wrong with the experiment but there were two major errors in the way it was carried out:

- the plant was operated below 20% output;
- the automatic shutdown equipment was isolated so that the experiment could be repeated.

When the temperature started to rise, it rose 100-fold in one second.

It is not clear why those in charge at Chernobyl ignored two major safety precautions. Perhaps they did not fully understand the hazards of the actions they took. Perhaps the importance of carrying out the tests — there would not be another opportunity for a year — had given them the impression that the safety rules could be relaxed (see Section 3.5, page 65). Whatever the reason, the most effective way of preventing similar accidents in the future is to change the design. The reactors should be redesigned so that they are not unstable at low output and it should be impossible to switch off the protective equipment. While we should try to persuade people to follow the rules — by explaining the reasons for them and by checking up to see that they are followed — when the results of not following the rules are serious we should not rely on this alone and should design safer plants[8,9].

Another fundamental error at Chernobyl was a failure to ask what would occur if the experiment was not successful. Before every experiment or change we should list possible outcomes and their effects and decide how they will be handled.

5.2 Accidents due to non-compliance by operators

These are also deliberate decisions, but easily made. When operators have to carry out routine tasks, and an accident is unlikely if the task, or part of it, is omitted or shortened, the temptation to take short cuts is great. It is done once, perhaps because the operator is exceptionally busy or tired, then a second time and soon becomes custom and practice. Some examples follow.

5.2.1 No-one knew the reason for the rule

Smoking was forbidden on a trichloroethylene (TCE) plant. The workers tried to ignite some TCE and found they could not do so. They decided that it would be safe to smoke. No-one had told them that TCE vapour drawn through a cigarette forms phosgene.

5.2.2 Preparation for maintenance

Many supervisors find permit-to-work procedures tedious. Their job, they feel, is to run the plant, not fill in forms. There is a temptation, for a quick job, not to bother. The fitter is experienced, has done the job before, so let's just ask him to fix the pump again.

They do so, and nothing goes wrong, so they do so again. Ultimately the fitter dismantles the wrong pump, or the right pump at the wrong time, and there is an accident. Or the fitter does not bother to wear the proper protective clothing requested on the permit-to-work. No-one says anything, to avoid unpleasantness; ultimately the fitter is injured.

To prevent these accidents, and most of those described in this chapter, a three-pronged approach is needed:

(1) We should try to convince people that the procedure — in this case a permit-to-work procedure — is necessary, preferably by describing accidents that have occurred because there was no adequate procedure or the procedure was not followed. Discussions are better than lectures or reports and the Institution of Chemical Engineers' safety training packages on *Safer Maintenance* and *Safer Work Permits* [4] can be used to explain the need for a permit-to-work system.

The discussion leader should outline the accident and then let the group question him to find out the rest of the facts. The group should then say what *they think* should be done to prevent the accident happening again.

(2) We should make the system as easy to use as possible and make sure that any equipment needed, such as locks and chains and slip-plates, is readily

available. Similarly, protective clothing should be as comfortable to wear as possible and readily available. If the correct method of working is difficult or time-consuming then an unsafe method will be used.

(3) Managers should check from time to time that the correct procedures are being followed. A friendly word the first time someone takes a short cut is more effective than punishing someone after an accident has occurred.

If an accident is the result of taking a short cut, it is unlikely that it occurred the first time the short cut was taken. It is more likely that short-cutting has been going on for weeks or months. A good manager would have spotted it and stopped it. If he does not, then when the accident occurs he shares the responsibility for it, legally and morally, even though he is not on the site at the time.

As already stated, a manager is not, of course, expected to stand over his team at all times. But he should carry out periodic inspections to check that procedures are being followed and he should not turn a blind eye when he sees unsafe practices in use (see Section 13.2, page 223).

The first step down the road to a serious accident occurs when a manager turns a blind eye to a missing blind.

5.2.3 An incident on a hydrogen/oxygen plant

An explosion occurred on a plant making hydrogen and oxygen by the electrolysis of water. As a result of corrosion some of the hydrogen had entered the oxygen stream.

Both streams were supposed to be analysed every hour. After the explosion, factory inspectors went through old record sheets and found that when conditions changed the analytical results on the sheets changed at once, although it would take an hour for a change to occur on the plant. It was obvious that the analyses were not being done.

The management had not noticed this and had failed to impress on the operators the importance of regular analyses and the results of not detecting hydrogen in the oxygen[5] (see also Section 4.2.2, page 82).

Engineering solutions, when possible, are usually better than reliance on manual tests and operator action and the official report recommended automatic monitoring and shutdowns (but see Section 7.5.2, page 143).

5.2.4 An example from the railways

The railways — or some of them — realized early on that when possible designers should make it difficult or impossible for people to bypass safety equipment. Engine drivers were tempted to tamper with the relief valves on early locomotives. As early as 1829, before their line was complete, the Liverpool and Manchester Railway laid down the following specification for their engines:

'There must be two safety valves, one of which must be completely out of reach or control of the engine man.' (See Figure 14.39, page 253.)

Incidentally, contrary to popular belief, very few locomotive boilers blew up because the driver tampered with the relief valve — a view encouraged by locomotive superintendents. Most explosions were due to corrosion (preventable by better inspection or better design), poor maintenance or delaying maintenance to keep engines on the road. Many of the famous locomotive designers were abysmal maintenance engineers[6].

5.2.5 Taking a chance

A man went up a ladder onto a walkway and then climbed over the handrails onto a fragile flat roof which was clearly labelled. He tried to stand only on the cross girders but slipped off onto the sheeting and fell through into the room below.

The accident was put down to 'human failing', the failure of the injured man to follow the rules. When the site of the accident was visited, however, it was found that the foot of the ladder leading to the roof was surrounded by junk, as shown in Figure 5.2 on page 106. It shows that too little attention was paid to safety in the plant and that in 'taking a chance' the injured man was following the example set by his bosses. They were, to some extent, responsible for the accident.

This incident also shows the importance, in investigating an accident, of visiting the scene and not just relying on reports.

5.2.6 Jobs half done

A particular form of non-compliance is to leave a job half-done. This may be harder to spot than a job omitted entirely.

The monthly test of a fire alarm system showed that the fire station was unable to receive signals from a number of the heat detectors. Three weeks earlier a new data terminal unit had been installed. The installers checked that the unit could send signals to the fire station. However, they did not check that the fire station was also able to receive signals. It could not do so[14].

A woman collected her car from a garage after repair of some accident damage. On the short journey home she had to make several turns. On each occasion she was hooted by other cars. When she got home and looked for the reason she found that the front indicator lights were connected wrongly:

when she indicated that she wanted to turn left, the right front indicator flashed. When she complained, the mechanic said he always checked indicator lights to make sure that they were connected correctly. However, he had checked the rear lights but not the front ones.

5.2.7 Non-compliance can be better than compliance

If instructions are wrong, impracticable or simply unwise, non-compliance may prevent an accident or further the objectives of the organization. For example, well-meaning store managers often estimate the money tied up by 'squirrels' and issue instructions such as, 'It is forbidden to draw material from store in anticipation of a possible future requirement.' The rule is often ignored, to the organization's advantage. Suppose a new fitter asks, 'What shall I do? There are no size 4 widgets in the store.' He may be told, 'Ask Fred. He keeps a few in his locker.' In other organizations the job is bodged or delayed (or bodging and delay start when Fred retires). As stated earlier, there is a fuzzy border between non-compliance and initiative.

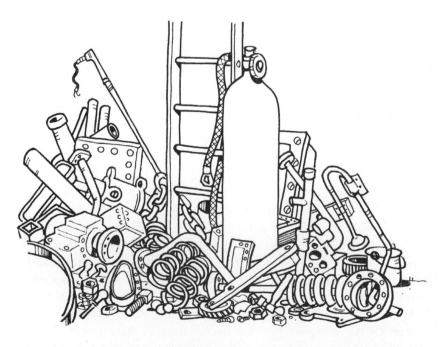

Figure 5.2 Foot of the ladder leading to a fragile roof. Did this scene encourage a man to take a chance?

5.3 Actions to improve compliance

In case anyone is tempted by motivational appeals, perhaps I should say that little or no improvement will result from generalized exhortations to people to work safely, follow the rules, be responsible or otherwise avoid sin.

Swain writes:

'Motivational appeals have a temporary effect because man adapts. He learns to tune out stimuli which are noise, in the sense of conveying no information. Safety campaigns which provide no useful, specific information fall into the "noise" category. They tend to be tuned out.'

And '... a search of the literature of industrial psychology has failed to show a single controlled experiment on the real effectiveness of safety motivation campaigns.'[7]

Swain is discussing general exhortations. Information and advice on specific problems can, of course, be effective.

To improve compliance with instructions we should first make sure that they are clear, up to date, easy to read, and contain the right amount of detail, as discussed in Sections 3.1 and 3.7 (pages 48 and 67). If they are too complex people will not refer to them and will simply follow custom and practice. Complex rules are often written to protect the writer rather than help the reader. Table 5.1 (page 108) gives some guidance on the degree of detail required.

We should, whenever possible, involve people in the preparation of the instructions they will be expected to follow. We should discuss new instructions with them and also the reasons for them (see Section 5.2.1, page 103). We may find that it is impossible to follow the instructions or that a better method of doing the job is possible! Reference 13 describes a formal procedure for involving operating staff in the development of good practice. Regular checks should then be made to make sure that instructions are followed. As already mentioned, a friendly word before an accident is more effective than a reprimand afterwards.

People are more influenced by their peers than by their bosses. If the group as a whole supports safe methods of working they may be able to influence people who carry out unsafe acts, either occasionally (most of us) or regularly. People can be trained to spot unsafe practices and to speak tactfully to those responsible. These behavioural science techniques can be very effective in preventing accidents caused by, for example, not wearing the correct protective clothing, using the wrong tool for a job, lifting incorrectly or leaving junk lying around[15].

Table 5.1 Decision aid to determine level of support required (based on a table supplied by A.G. Foord and W. Gulland, private communication)

Task criticality	Low			Medium			High		
Task familiarity	Frequent	Infrequent	Rare	Frequent	Infrequent	Rare	Frequent	Infrequent	Rare
Task complexity:									
Low	NWI	NWI	NWI	NWI	NWI	JA	NWI	NWI	JA
Medium	NWI	NWI	NWI	NWI	NWI	JA	NWI	JA	SBS
High	NWI	NWI	JA	NWI	JA	SBS	JA	JA	SBS

NWI = No Written Instruction required
JA = Job Aid required (for example, check-list / memory aid)
SBS = Step-By-Step instruction required

They cannot, of course, have much effect on accidents due to poor training or instructions, slips or lapses of attention, lack of ability, non-compliance by managers or poor process or engineering design. Companies have achieved low lost-time and minor accident rates by behavioural safety training, felt that their safety record was good and then had a serious fire or explosion because an aspect of the technology was not fully understood or the precautions necessary had been forgotten.

Another limitation of behavioural training is that its effects can lapse if employees feel that their pay or working conditions (what Herzberg calls hygienic factors; see Appendix 1) are below expectations.

Behavioural safety training is effective in preventing accidents that may injure the person carrying them out, or his immediate fellow workers. It has been much less effective in preventing incidents such as the two following. Both occurred in a factory that had introduced behavioural training and achieved major reductions in the accident rate.

There were two filters in parallel — one working, one spare. The working one choked and the spare was put on line. No-one reported this or had the choked filter replaced. When the second filter choked, the unit had to be shut down. In many cases, when we neglect a safety measure, we get away with it. In this case, the result was inevitable.

A fitter was asked to dismantle a valve of an unfamiliar type. As a result there was a serious leak of a hazardous chemical. The pipeline he was working on had not been isolated correctly, but if he had dismantled the valve in the correct way the leak could have been controlled. A more safety-conscious man would have got a spare valve out of store and dismantled that first. However, this would have held up the job. The manager would have accepted the delay but the culture of the workplace was 'get stuck in and get the job done'. Also, it would have meant admitting to his fellow workers that he did not know how the valve was constructed. The behavioural training was not powerful enough to overcome these two aspects of the culture and trainers had not recognized the problem.

5.4 Alienation

Failure to follow instructions or recognized procedures can be due to alienation. In extreme cases an employee may deliberately damage products or equipment. From his point of view his decisions are not wrong as his objectives are not the same as those of the managers. In such cases advice is needed from an employee relations expert rather than a safety professional (see Appendix 1). An example from history rather than industry will show how failure to take the realities of human nature into account can wreck a project. The 12th century Crusader Castle of Belvoir in Israel, overlooking the Jordan Valley, appears impregnable. There are cliffs on three sides, a deep dry moat and several defensive walls. How, the visitor wonders, did such a castle ever fall to an enemy?

Many of the defenders were mercenaries. The innermost fortification was reserved for the Crusader knights. The mercenaries knew that they were ultimately expendable and when the going got rough they were tempted to change sides[10]. The safety of a project depends on the quality of the procedures as well as the quality of the hardware. Mercenaries are contractors. If we treat contractors less favourably than employees, we cannot expect the same commitment from them.

The case for and against punishment and prosecution is discussed in Section 13.3, page 227.

5.5 Postscript

Describing the attitude of the Amish, an American Puritan Sect, towards children, Huntingdon writes[11]:

'When they disobey it is generally because the parents have not stated the instructions in the way they should, have expected too much of the child, have been inconsistent, or have not set a good example.'

References in Chapter 5

1. Howard, H.B., 1984, *Plant/Operations Progress*, 3(3): 147.
2. *Hansard*, 8 May 1980.
3. UK Atomic Energy Authority Systems Reliability Service, Warrington, UK, 1974, *Report of Annual General Meeting*, page 74.
4. Safety training packages 028 *Safer Work Permits* and 033 *Safer Maintenance* (Institution of Chemical Engineers, Rugby, UK).
5. Health and Safety Executive, 1976, *The Explosion at Laporte Industries Limited on 5 April 1975* (HMSO, London, UK).
6. Hewison, C.H., 1983, *Locomotive Boiler Explosions* (David and Charles, Newton Abbot, UK).
7. Swain, A.D., A work situation approach to improving job safety, in Widner, J.T. (editor), 1973, *Selected Readings in Safety*, page 371 (Academy Press, Macon, Georgia, USA).
8. Kletz, T.A., 2001, *Learning from Accidents*, 3rd edition, Chapter 12 (Butterworth-Heinemann, Oxford, UK).

9. Gittus, F.H. *et al.*, 1987, *The Chernobyl Accident and its Consequences* (HMSO, London, UK).
10. Murphy-O'Connor, J., 1998, *The Holy Land*, 4th edition, page 179 (Oxford University Press, Oxford, UK).
11. Huntingdon, G.E., 1990, *Natural History*, No. 4, page 94.
12. Anon, quoted by Long, J., 1983, *Risk and Benefit — A Report and Reflection on a Series of Consultations* (St George's House, Windsor, UK).
13. Embrey, D., 1998, Carman: A systematic approach to risk reduction by improving compliance to procedures, *Hazards XIV — Cost Effective Safety, Symposium Series No. 144* (Institution of Chemical Engineers, Rugby, UK).
14. *Operating Experience Weekly Summary*, 1998, No 98–48, page 2 (Office of Nuclear and Facility Safety, US Department of Energy, Washington, DC, USA).
15. Sellers, G. and Eyre, P., 2000, The behaviour-based approach to safety, *Hazards XV: The Process, its Safety and the Environment — Getting it Right!, Symposium Series No. 147* (Institution of Chemical Engineers, Rugby, UK).

Accidents that could be prevented by better management

6

'When did you last see the senior members of your ...
team and re-emphasise the policy to them, ask how they
were getting on, what problems they were meeting, what
success they had had ...?'
Lord Sieff (former chairman of Marks & Spencer)[1]

The last four chapters have described four sorts of accidents: those due to slips or lapses of attention; those due to poor training or instruction (mistakes); those due to a mismatch between the job and the ability of the person asked to do it; and those due to non-compliance with rules or accepted practice. The accidents described in this chapter are not due to a fifth type of error. They are due to the failure of senior managers to realize that they could do more to prevent accidents. They are thus mainly due to lack of training, but some may be due to lack of ability and a few to a deliberate decision to give safety a low priority. As stated in Section 5.1, these accidents are sometimes said to be due to organizational failures but organizations have no minds of their own. Someone has to change the culture of the organization and this needs the involvement, or at least the support, of senior people. During my time as a safety adviser with ICI Petrochemicals Division, my team did bring about a change in the Division's attitude to safety, but we could only do so because everyone knew that the Board supported our activities.

If output, costs, efficiency or product quality require attention, senior managers identify the problems, agree actions and ask for regular reports on progress. This approach is rarely seen where safety is concerned. Although senior managers repeatedly say that safety is important, they rarely show the level of detailed interest that they devote to other problem areas. Many look only at the lost-time accident rate but in many companies today all it measures is luck and the willingness of injured employees to return to work. If senior managers comment on the lost-time accident rate and nothing else they give staff the impression that they are not interested in the real problems. The results are the opposite of those intended: everyone thinks, 'If the senior managers are not interested, the problems can't be important.'

If output, costs, efficiency and product quality fall, the results are soon apparent and obviously require immediate action. In contrast, if safety standards fall, it may be a long time before a serious accident occurs. The fall in standards is hidden or latent. This may be one reason why safety is given insufficient detailed attention and instead exhortation to work safely replaces consideration of the real problems. Another reason is that senior managers want a single measure of safety performance. The lost-time accident rate seems to supply this, but does not. Other possible measures are discussed in Section 6.8 (page 125).

At this point in earlier chapters I summarized the actions needed to prevent accidents similar to those described in the pages that followed. When the accidents described in this chapter were the subject of official reports, the recommendations were quite clear and can be summed up by the words of the report on the Clapham Junction railway accident (Section 6.5, page 118): ' … a concern for safety which is sincerely held and repeatedly expressed but, nevertheless, is not carried through into action, is as much protection from danger as no concern at all.' Unfortunately, most managers do not think it applies to them and who will tell them otherwise? Authors of company accident reports, such as those summarized in Sections 6.1 and 6.2 (pages 114 and 115), are unlikely to suggest that their directors could do better and most directors do not attend safety conferences. They send the safety officer instead.

In earlier chapters I tried to make it clear that blame is rarely an appropriate action after an accident. Instead we should look for ways of reducing opportunities for error, of improving training or instruction and so on. The same applies to managers' errors. Few are due to a deliberate decision to ignore risks. This point is worth emphasizing because writers who are ready to excuse the errors of ordinary people, even those of criminals, and prefer to tackle underlying causes rather than look for culprits, seem to take a different view when managers fail. For example, two academics have written a book called *Toxic Capitalism: Corporate Crime and the Chemical Industry*. A reviewer summarized its message as 'corporations are unable see anything other than safety as an avoidable cost' and 'the safety professional is helpless against the higher pressures of corporate strategies that play down their role'[10]. I do not think this applies to the majority of companies though it obviously applies to a few cowboys. Directors are not supermen. They fail as often as the rest of us and for much the same reasons: not wickedness, not a deliberate decision to ignore hazards, not slips and lapses of attention as they have time to check their work, but ignorance, which they do not recognize, occasionally incompetence, and all the other weaknesses of human nature that affect the rest of us, such as repeatedly postponing jobs that can wait until tomorrow in order to do something that must be done today and ignoring matters peripheral to their core responsibilities.

6.1 An accident caused by insularity[2]

An explosion, which killed four men, occurred in a plant which processed ethylene at high pressure. A leak from a badly-made joint was ignited by an unknown cause. After the explosion many changes were made to improve the standard of joint-making. The training, tools and inspection were all improved.

Poor joint-making had been tolerated for a long time before the explosion because all sources of ignition had been eliminated and so leaks could not ignite, or so it was believed. The plant was part of a large group but the individual parts of it were technically independent. The other plants in the group had never believed that leaks of flammable gas will not ignite. They knew from their own experience that sources of ignition are liable to turn up, even though we do everything we can to remove known sources, and therefore strenuous efforts must be made to prevent leaks. Unfortunately the managers of the ethylene plant had hardly any technical contact with the other plants, though they were not far away; handling flammable gases at high pressure was, they believed, a specialized technology and little could be learnt from those who handled them at low pressure. The factory was a monastery, a group of people isolating themselves from the outside world. The explosion blew down the monastery walls.

If the management of the plant where the explosion occurred had been less insular and more willing to compare experiences with other people in the group, or if the managers of the group had allowed the component parts less autonomy, the explosion might never have occurred. It is doubtful if the senior managers of the plant or the group ever realized or accepted this or discussed the need for a change in policy. The leak was due to a badly-made joint and so joints must be made correctly in future. No expense was spared to achieve this aim but the underlying weaknesses in the management system went largely unrecognized. However, some years later, during a recession, the various parts of the group were merged.

Similarly, the official report on the King's Cross fire (see Section 6.3, page 116) said that there was 'little exchange of information or ideas between departments and still less cross-fertilisation with other industries and outside organisations' (Paragraph 9).

6.2 An accident due to amateurism[3]

This accident is somewhat similar to the last one. A chemical company, part of a large group, made all its products by batch processes. They were acknowledged experts and had a high reputation for safety and efficiency. Sales of one product grew to the point that a large continuous plant was necessary. No-one in the company had much experience of such plants so they engaged a contractor and left the design to him. If they had consulted the other companies in the group they would have been told to watch the contractor closely and told some of the points to watch.

The contractor sold them a good process design but one that was poor in other ways; the layout was very congested, the drains were open channels and the plant was a mixture of series and parallel operation.

When some of the parallel sections of the plant were shut down for overhaul other sections were kept on line, isolated by slip-plates. One of the slip-plates was overlooked. Four tonnes of a hot, flammable hydrocarbon leaked out of an open end and was ignited by a diesel engine used by the maintenance team. Two men were killed and the plant was damaged. The congested design increased the damage and the open drainage channels allowed the fire to spread rapidly.

The immediate cause of the fire was the missing slip-plate. The foreman who decided where to put the slip-plates should have followed a more thorough and systematic procedure. But the underlying cause was the amateurism of the senior management and their failure to consult the other companies in the group. If they had consulted them they would have been told to watch the contractor closely; to divide the plant into blocks with breaks in between, like fire breaks in a forest; to put the drains underground; and, if possible, not to maintain half the plant while the other half is on line. If it is essential to do so then they would have been told to build the two halves well apart and to plan well in advance, at the design stage, where slip-plates should be inserted to isolate the two sections. It should not be left for a foreman to sort out a few days before the shutdown.

I doubt if it ever occurred to the senior managers of the company or group that their decisions had led to the accident, though they fully accepted that they were responsible for everything that went on. No changes in organization were made immediately but a few years later responsibility for the continuous plant was transferred to another company in the group.

People can be expert in one field but amateurs in another into which their technology has strayed.

6.3 The fire at King's Cross railway station

The fire at King's Cross underground station, London, in 1987, killed 31 people and injured many more. The immediate cause was a lighted match, dropped by a passenger on an escalator, which set fire to an accumulation of grease and dust on the escalator running track. A metal cleat which should have prevented matches falling through the space between the treads and the skirting board was missing and the running tracks were not cleaned regularly.

No water was applied to the fire. It spread to the wooden treads, skirting boards and balustrades and after 20 minutes a sudden eruption of flame occurred into the ticket hall above the escalator. The water spray system installed under the escalator was not actuated automatically and the acting inspector on duty walked right past the unlabelled water valves. London Underground employees, promoted largely on the basis of seniority, had little or no training in emergency procedures and their reactions were haphazard and unco-ordinated.

Although the combination of a match, grease, dust and a missing cleat were the immediate causes of the fire, an underlying cause was the view, accepted by all concerned, including the highest levels of management, that occasional fires on escalators and other equipment were inevitable and could be extinguished before they caused serious damage or injury. From 1958 to 1967 there were an average of 20 fires per year, called 'smoulderings' to make them seem less serious. Some had caused damage and passengers had suffered from smoke inhalation but no-one had been killed. The view thus grew that no fire could become serious and they were treated almost casually. Recommendations made after previous fires were not followed up. Yet escalator fires could have been prevented, or reduced in number and size, by replacing wooden escalators by metal ones, by regular cleaning, by using non-flammable grease, by replacing missing cleats, by installing smoke detectors which automatically switched on the water spray, by better training in fire-fighting and by calling the fire brigade whenever a fire was detected, not just when it seemed to be getting out of control. There was no defence in depth.

The management of safety in London Underground was criticized in the official report[4]. There was no clear definition of responsibility, no auditing, no interest at senior levels. It seems that London Underground ran trains very competently and professionally but was less interested in peripheral matters such as stations. (In the same way, many process managers give service lines less than their fair share of attention and they are involved in a disproportionate number of incidents.) (See also the quotation at the end of Section 6.1, page 114.)

6.4 The Herald of Free Enterprise[5]

In 1987 the cross-Channel roll-on/roll-off ferry *Herald of Free Enterprise* sank, with the loss of 186 passengers and crew, soon after leaving Zeebrugge in Belgium, *en route* for Dover. The inner and outer bow doors had been left open because the assistant bosun, who should have closed them, was asleep in his cabin and did not hear an announcement on the loudspeakers that the ship was ready to sail. However, the underlying causes were weaknesses in design and poor management. 'From top to bottom the body corporate was infected with the disease of sloppiness.' The official reports, from which this and other quotations are taken, shows that if we look below the obvious causes and recommendations we find ways of improving the management system.

Managerial 'sloppiness' occurred at all levels. The officer in charge of loading did not check that the assistant bosun was on the job. He was unable to recognize him as the officers and crew worked different shift systems. There was pressure to keep to time and the boat was late. Before sailing the captain was not told that everything was OK; if no defects were reported, he assumed it was. There was no monitoring system. Responsibility for safety was not clear; one director told the Court of Inquiry that he was responsible; another said no-one was. '… those charged with the management of the Company's ro-ro fleet were not qualified to deal with many nautical matters and were unwilling to listen to their Masters, who were well-qualified.' In particular, requests for indicator lights to show that the doors were shut were not merely ignored but treated as absurd. Complaints of overloading were brushed aside. After the disaster they were fitted at a cost of only £500 per ship[11]. Ships were known to have sailed before with open doors but nothing was done.

In addition to the detailed errors in design, such as the absence of indicator lights, there was a more serious and fundamental error. In the holds of most ships there are partitions to confine any water that enters. On most ro-ro ferries the partitions were omitted so that vehicles had a large uninterrupted parking space. Any water that entered could move to one side, as the ship rolled, and make it unstable. Safety depended on keeping water out. Defence in depth was lost.

The Certificates of Competency of the captain and first officer were suspended but, 'the underlying or cardinal faults lay higher up in the Company. The Board of Directors did not appreciate their responsibility for the safe management of their ships … They did not have any proper comprehension of what their duties were.' The Directors, however, seem to have continued in office. In fact, six months after the tragedy the Chairman of the holding company was quoted in the press as saying, 'Shore based

117

management could not be blamed for duties not carried out at sea.'[6] He had spent most of his career in property management and may not have realized that managers in industry accept responsibility for everything that goes on, even though it is hundreds of miles from their offices. Sooner or later people fail to carry out routine tasks and so a monitoring or audit system should be established. This may have been difficult on ships in the days of Captain Cook, when ships on long voyages were not seen for months, even years, but today a ferry on a short sea crossing is as easy to audit as a fixed plant.

The Zeebrugge report should be read by any director or senior manager who thinks that safety can be left to those on the job and that all they need do is produce a few expressions of goodwill.

6.5 The Clapham Junction railway accident

The immediate causes of this 1989 accident, in which 35 people were killed and nearly 500 injured, were repeated non-compliances of the type discussed in Chapter 5 and a slip of the type discussed in Chapter 2, but the underlying cause was failure of the managers to take, or even to see the need to take, the action they should have taken.

The non-compliances, errors in the way wiring was carried out by a signalling technician (not cutting disused wires back, not securing them out of the way and not using new insulation tape on the bare ends), were not isolated incidents; they had become his standard working practices (Paragraph 8.2). (Paragraph numbers refer to the official report[7].) 'That he could have continued year after year to continue these practices, without discovery, without correction and without training illustrates a deplorable level of monitoring and supervision within BR (British Rail) which amounted to a total lack of such vital management actions' (Paragraph 8.4).

In addition the technician made 'two further totally uncharacteristic mistakes' (disconnecting a wire at one end only and not insulating the other, bare end), perhaps because 'his concentration was broken by an interruption of some sort' (Paragraph 8.22) and because of 'the blunting of the sharp edge of close attention which working every day of the week, without the refreshing factor of days off, produces' (Paragraph 8.26). 'Any worker will make mistakes during his working life. No matter how conscientious he is in preparing and carrying out his work, there will come a time when he will make a slip. It is those unusual and infrequent events that have to be guarded against by a system of independent checking of his work' (Paragraph 8.27). Such a system, a wire count by an independent person or even by the person who did the work, was lacking.

The supervisor was so busy working himself that he neglected his duties as supervisor (Paragraph 8.34).

The original system of checking was a three-level one; the installer, the supervisor and a tester were supposed to carry out independent checks (Paragraph 8.44). In such a system people tend after a while to neglect checks as they assume that the others will find any faults. Asking for too much checking can increase the number of undetected faults (a point not made in the official report). See Section 7.9 (page 147).

Other errors contributing to the accident were:

- Failures in communication: instructions were not received or not read (Paragraphs 8.30–41, 9.7, 9.23 and 9.44). New instructions should be explained to those who will have to carry them out, not just sent to them through the post (see Section 3.7, page 67).
- A failure to learn the lessons of the past: similar wiring errors had been made a few years before though with less serious results (Paragraphs 9.50–53 and 9.59).
- A failure to employ suitable staff: the testing and commissioning signal engineer 'was doing a job which he had really no wish to do at a place where he had no wish to be. If he had little liking for the job he had less enthusiasm' (Paragraph 9.5).

  The supervisor above the immediate supervisor of the technician turned a blind eye to the fact that good practice was not being followed. He had been 'pitched in the twilight of his career into work that was foreign to him' (Paragraph 16.18).
- A failure to follow up: one manager identified the problems in the signalling department (poor testing) and arrived at a solution (new instructions and an élite group of testers with a new engineer in charge) but then he turned his attention to other things and made the assumption that the solution would work and the problem would go away (Paragraphs 16.64–66).

All these errors add up to an indictment of the senior management who seem to have had little idea what was going on. The official report makes it clear that there was a sincere concern for safety at all levels of management but there was a 'failure to carry that concern through into action. It has to be said that a concern for safety which is sincerely held and repeatedly expressed but, nevertheless, is not carried through into action, is as much protection from danger as no concern at all' (Paragraph 17.4).

6.6 Piper Alpha

In 1988 the explosion and fire on the Piper Alpha oil platform in the North Sea claimed 167 lives. The immediate cause was a poor permit-to-work system and poor handover between shifts but the underlying cause reinforces the message of the other incidents described in this chapter. To quote from the official report[9]:

'I do not fault Occidental's policy or organisation in relation to matters of safety. However, … I have had to consider a number of shortcomings in what existed or took place on Piper. This calls in question the quality of Occidental's management of safety, and in particular whether the systems which they had for implementing the company safety policy were being operated in an effective manner.'

'No senior manager appeared to "own" the problem [blockages in water deluge lines] and pursue it to an early and satisfactory conclusion.'

The top men in Occidental were not hard-nosed and uncaring people interested only in profit and unconcerned about safety. They said and believed all the right things; they said that safety was important but they did not get involved in the precise actions required, see that they were carried out and monitor progress.

'Safety is not an intellectual exercise to keep us in work. It is a matter of life and death. It is the sum of our contributions to safety management that determines whether the people we work with live or die. On Piper Alpha they died.' (Brian Appleton, Technical assessor to the public enquiry.)

6.7 What more can senior managers do?

I suggest at the beginning of this chapter that senior managers should identify the major safety problems, agree actions and follow up to see that actions have been taken. The problems differ from company to company and from time to time but the following are widespread.

'I've appointed a good man as safety adviser and given him the resources he wants. What more can I do?' Here are some suggestions.

6.7.1 The need for user-friendly designs

As discussed in Chapter 2, people are actually very reliable but there are many opportunities for error in the course of a day's work. When hazardous materials are handled the lowest error rates obtainable, from equipment as well as people, may be too high.

Increasingly, therefore, the process industries are trying to design plants and equipment that are inherently safer or user-friendly — that is, they can withstand human error or equipment failure without serious effects on safety (and output and efficiency). The following are some of the ways in which plants can be made more user-friendly.

- Use so little hazardous material that it does not matter if it all leaks out (intensification). 'What you don't have, can't leak.'
- Use a safer material instead (substitution). Use hazardous materials in the least hazardous form (attenuation).
- Simplify the design. Complexity means more opportunities for human error and more equipment that can fail.
- Limit effects, not by adding on protective equipment but by equipment design, by changing reaction conditions, by limiting the energy available or by eliminating hazardous phases, equipment or operations.
- Avoid knock-on or domino effects.
- Make incorrect assembly impossible.
- Make the status of equipment (open or closed, on-line or off-line) clear.
- Use equipment that can tolerate poor operation or maintenance.

These may seem obvious but until after the explosion at Flixborough in 1974 (see Section 3.6, page 66) little or no thought was given to ways of reducing the inventory of hazardous material in a plant. People accepted whatever inventory was required by the design, confident of their ability to keep the lions under control. Flixborough weakened the public's confidence in that ability and ten years later the toxic release at Bhopal (over 2000 killed) almost destroyed it. Now many companies are coming round to the view that they should see if they can keep lambs instead of lions. In many cases they can do so[12]. Such plants are often cheaper as well as safer as they contain less added-on protective equipment which is expensive do buy and maintain. In addition, if we can intensify, then we need smaller equipment and the plant will be correspondingly cheaper.

Senior managers should ask for regular reports on progress in reducing the inventories in new and existing plants and ask how new designs compare with old ones.

6.7.2 The need to see that the right procedures are used and the right mix of knowledge and experience is available

To achieve the user-friendly designs just described we need to consider alternatives systematically during the early stages of design. Very often safety studies do not take place until late in design when all we can do is control hazards by adding-on additional protective equipment or procedures. Such a change will not come about unless senior managers actively encourage it. The same applies to safety management systems and specific procedures such as those for controlling modifications and preparing equipment for maintenance.

There has been an explosion of interest in safety management systems in recent years and no topic is more popular at conferences. Some recent incidents have left me with an uneasy feeling that some managers believe that a good safety management system is all they need for safe performance. All it can do, however, is ensure that people's knowledge and experience are applied systematically and thus reduce the chance that something is missed. If the staff lack knowledge and experience then the system is an empty shell. People will go through the motions but the output will be poor. This is a particular danger at the present time when companies are reducing manning and the over-fifties are looked upon as expenses to be eliminated rather than assets in which thirty years' salary has been invested. *Senior managers should systematically assess the levels of knowledge and experience needed and ensure that they are maintained.*

6.7.3 The need for more thorough investigation of accidents

When an accident has occurred the facts are usually recorded but, as the accidents described in this chapter show, we often draw only superficial conclusions from them. We identify the immediate technical causes but we do not look for the underlying weaknesses in the management system.

Petroski writes[13]:

'Case histories of failures often tend to focus in excruciating detail on the specifics of the failed design and on the failure mechanism. Such a concentrated view often discourages all but the most narrow of specialists from pursuing the case history to any degree, for the lessons learned and conclusions drawn tend to be so case-specific as to seem hardly relevant to anything but a clone of the failed design.'

Accident investigation is rather like peeling an onion or dismantling a Russian doll. Beneath each layer of cause and effect are other layers. The outer ones deal with the immediate technical causes, the inner ones with ways of avoiding the hazards and with weaknesses in the management system. We tend to pick on the one, often the immediate technical cause, that is nearest to our current interests. Some accident reports tell us more about the interest and prejudices of the writer than about the most effective methods of prevention.

It is often difficult for a junior manager, closely involved in the detail, to see beyond the immediate technical causes. If someone from another plant is included in the investigating team he may see them more clearly. *Senior managers should not accept reports that deal only with immediate causes.*

6.7.4 The need to learn and remember the lessons of the past

Even when accidents are investigated thoroughly the lessons are often forgotten.

In 1943 an elderly Aborigine led his group on a six-month trek to escape a drought in Western Australia.

His first goal was a water hole at the extreme north-western corner of the tribal territory, which he had visited only once in his youth, more than half a century earlier. When the resources there started to fail, he led them westwards again, through territory known to him only through the verses of a song cycle sung at totemic ceremonies and depicting the legendary wanderings of 'ancestral beings'. The trek led on through a sequence of more than 50 water holes, with the only additional clues to the route being occasional marks left by earlier movements of peoples. The little band finally emerged ... more than 360 miles from where they started[14].

The Australian Aborigines lack (or did at the time) literacy, computers and management consultants. Yet they did better than many high technology companies do today. All too often we forget the lessons of the past and the same accidents happen again. However, the Aborigines have one advantage that we lack: people stay in the same group for a long time. In a typical factory it is often hard to find anyone who has been on the same plant for more than ten years. After a while people move, taking their memories with them, and the lessons learnt after an accident are forgotten. Someone, keen to improve efficiency, a very desirable thing to do, asks why we are following a time-consuming procedure or using inconvenient equipment. No-one knows, so the procedure is abandoned or the equipment removed and the accident it was introduced to prevent happens again. The present enthusiasm for downsizing makes the problem worse; we can manage without advice and experience until we fall into the trap that no-one knew was there, except the people who retired early.

The following actions can help us remember the lessons of the past. If we have paid the high price of an accident we should at least turn it into a learning experience[15].

- Include in every instruction, code and standard a note on the reason why it was introduced and the accidents that would not have occurred if it had been followed.

- Describe old accidents as well as recent ones in safety bulletins and newsletters and discuss them at safety meetings. Giving the message once is not enough.
- Follow up at regular intervals to see that the recommendations made after accidents are being followed, in design as well as operations.
- Remember that the first step down the road to the next accident occurs when someone turns a blind eye to a missing blind.
- Never remove equipment before we know why it was installed. Never abandon a procedure before we know why it was adopted.
- Include important accidents of the past in the training of undergraduates and company employees.
- Ask experienced people, before they leave or retire, to write down their know-how, especially the information that younger and less experienced people are not aware of.
- Read books that tell us what is old as well as magazines that tell us what is new.
- Devise better retrieval systems so that we can find, more easily than at present, details of past accidents, in our own and other companies, and the recommendations made afterwards. Existing accident databases (including the Institution of Chemical Engineers') are not being used as much as we hoped they would be. We consult them only when we suspect that there may be a hazard. If we don't suspect there may be a hazard, we don't look. The computer is passive and the user is active. The user has to ask the database if there is any information on, say, accidents involving non-return (check) valves. We need a system in which the user is passive and the computer is active. With such a system, if someone types the words 'non-return valve' (or perhaps even makes a diary entry that there is going to be a meeting on non-return valves), the computer will signal that the database contains information on these subjects and a click of the mouse will then display the data. As I type these words the spell-check and grammar-check programs are running in the background, drawing my attention to my (frequent) spelling and grammar errors. In a similar way, a safety database could draw attention to any subject on which it has data. Filters could prevent it repeatedly referring to the same hazard.

Another weakness of existing search engines is their hit or miss nature. We either get a hit or we don't. Suppose we are looking in a safety database to see if there are any reports on accidents involving the transport of sulphuric acid. Most search engines will either display them or tell us there are none. 'Fuzzy' search engines will offer us reports on the transport of other minerals acids or perhaps on the storage of sulphuric acid.

This is done by arranging keywords in a sort of family tree. If there are no reports on the keyword, the system will offer reports on its parents or siblings[16].

6.7.5 Management education

A survey of management handbooks[17] shows that most of them contain little or nothing on safety. For example, *The Financial Times Handbook of Management* (1184 pages, 1995) has a section on crisis management but 'there is nothing to suggest that it is the function of managers to prevent or avoid accidents'. *The Essential Manager's Manual* (1998) discusses business risk but not accident risk while *The Big Small Business Guide* (1996) has two sentences to say that one must comply with legislation. In contrast, the *Handbook of Management Skills* (1990) devotes 15 pages to the management of health and safety. Syllabuses and books for MBA courses and National Vocational Qualifications in management contain nothing on safety or just a few lines on legal requirements.

6.8 The measurement of safety

If the lost-time accident rate is not an effective measure of safety, what could we measure instead or as well? Here are some suggestions.

(1) A monthly summary of the cost of fires and other dangerous occurrences can draw attention to problem areas and their effect on profits. The costs can be divided into insured costs, for which precise figures are available, uninsured costs and business interruption costs.

(2) As suggested in Section 6.7.1 (page 120), an annual report on the progress made on reducing inventories of hazardous materials, both in process and in storage, can concentrate attention on one of the most effective ways of preventing large leaks.

(3) Many companies use the results of audits to compare one plant with another and one year with another. Reference 18 is an anthology of the methods used in various companies. Other companies prefer surveys of employees' perceptions and/or performance monitoring[19]. Compared with the lost-time accident rate and the costs of damage these methods try to detect falling standards before an accident occurs.

(4) The number of faulty permits-to-work found by routine inspection (not after incidents) and/or the number of trips and alarms found to be faulty. A good

practice is to display charts, similar to Figure 6.1, in each control building showing the status of all trips and alarms. Everyone entering the building can see their present and recent condition at a glance.

(5) Many accidents and dangerous occurrences are preceded by near misses, such as leaks of flammable liquids and gases that do not ignite. Coming events cast their shadows before. If we learn from these we can prevent many accidents. However, this method is not quantitative. If too much attention is paid to the *number* of dangerous occurrences rather than their lessons, or if numerical targets are set, then some dangerous occurrences will not be reported.

Week	1	2	3	4	5	6	7	8	9	10
A	OK				OK				*	
B		OK				OK				*
C			FR				OK			
D				OK				*		
E	OK				OK				*	
F						OK				*
G			OK				F			
H				RM				*		
I	OK				OK				*	
J		RP				OK				*

Key

*	= Test due	OK	= Test passed
RP	= Test refused by process team	F	= Test failed, awaiting repair
RM	= Test refused by maintenance team	FR	= Test failed, now repaired

Stars, squares etc of different colours stand out more clearly than letters

Figure 6.1 Charts similar to this one show when tests are due and the results of recent tests. A, B, C ... indicate the names or reference numbers of trips and alarms. Everyone entering the control room can see their status at a glance.

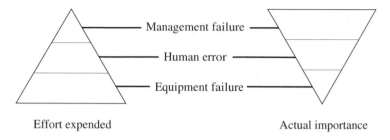

Effort expended   Actual importance

Figure 6.2 The effort expended on the causes of accidents and their relative importance

6.9 Conclusions

The left-hand side of Figure 6.2 (adapted from Reference 8) shows the relative effort devoted in the past to the prevention of equipment failure, the prevention of human error (as traditionally understood) and the prevention of management failures. The right-hand side show their relative importance.

'He saw the open and more superficial errors in persons and individuals but not the hidden and more profound errors in structures and systems. And because of that he sought the roots of Roman decline not in its empire but in its emperors, never recognizing the latter as but the former's personification.'

The Roman author Tacitus described by John D. Crossan[20] (I have changed *evil* to *errors*).

But who can change the empire but the emperor?

References in Chapter 6

1. Sieff, M., 1988, *Don't Ask the Price*, page 329 (Collins Fontana, London, UK).
2. Kletz, T.A., 2001, *Learning from Accidents*, 3rd edition, Chapter 4 (Butterworth-Heinemann, Oxford, UK).
3. Kletz, T.A., 2001, *Learning from Accidents*, 3rd edition, Chapter 5 (Butterworth-Heinemann, Oxford, UK).
4. Fennell, D. (Chairman), 1988, *Investigation into the King's Cross Underground Fire* (HMSO, London, UK).
5. Department of Transport, 1987, *MV Herald of Free Enterprise: Report of Court No. 8074: Formal Investigation* (HMSO, London, UK).
6. *Daily Telegraph*, 10 October 1987.
7. Hidden, A. (Chairman), 1989, *Investigation into the Clapham Junction Railway Accident* (HMSO, London, UK).

8. Batstone, R.J., 1987, *International Symposium on Preventing Major Chemical Accidents, Washington, DC, February 1987* (not included in published proceedings).
9. Cullen, W.D., 1990, *The Public Inquiry into the Piper Alpha Disaster*, Paragraphs 14.10 and 14.51 (HMSO, London, UK).
10. Gill, F., 1999, *Health and Safety at Work*, 21(1): 34. The book reviewed is by F. Pearce and S. Tombs and is published by Ashgave, Aldershot, UK.
11. Crainer, S., 1993, *Zeebrugge — Learning from Disaster*, page 78 (Herald Families Association, London, UK).
12. Kletz, T.A., 1998, *Process Plants: A Handbook for Inherently Safety Design*, 2nd edition (Taylor & Francis, Philadelphia, PA, USA).
13. Petroski, H., 1994, *Design Paradigms*, page 82 (Cambridge University Press, Cambridge, UK).
14. Brace, C.L., *Natural History*, February 1997, page 12.
15. Kletz, T.A., 1993, *Lessons from Disaster – How Organisations have no Memory and Accidents Recur* (Institution of Chemical Engineers, Rugby, UK).
16. Chung, P.W.H., Iliffe, R.E. and Jefferson, M., 1998, Integration of accident database with computer tools, *IChemE Research Event, Newcastle, UK, 7–8 April 1998.*
17. Buttolph, M., 1999, *Health and Safety at Work*, 21(7): 20, 21(8): 24 and 21(9): 29.
18. van Steen, J. (editor), 1996, *Safety Performance Measurement* (Institution of Chemical Engineers, Rugby, UK).
19. Petersen, D., 1996, *Safety by Objectives*, 2nd edition (van Nostrand Reinhold, New York, USA).
20. Crossan, J.D., 1995, *Who Killed Jesus?*, page 15 (Harper Collins, San Francisco, CA, USA).

The probability of human error

'The best of them is but an approximation, while the worst bears no relation whatever to the truth.'
E.C.K. Gonner[12]

This chapter lists some estimates that have been made of the probability of error. However, these data have several limitations.

First, the figures are estimates of the probability that someone will, for example, have a moment's forgetfulness or lapse of attention and forget to close a valve or close the wrong valve, press the wrong button, make a mistake in arithmetic and so on, errors of the type described in Chapter 2. They are not estimates of the probability of error due to poor training or instructions (Chapter 3), lack of physical or mental ability (Chapter 4), lack of motivation (Chapter 5) or poor management (Chapter 6). There is no recognized way of estimating the probabilities of such errors, but we can perhaps assume that they will continue in an organization at the same rate as in the past, unless there is evidence of change. One company has developed a system for carrying out an audit of the management, awarding marks under various headings and multiplying *equipment* failure rates by a factor between 0.1 and 100 derived from the results[13]. They assume that equipment failure rates vary over a range of 1000:1 depending on the quality of the management (see Section 7.10, page 149).

Second, the figures are mainly estimates by experienced managers and are not based on a large number of observations. Where operations such as soldering electric components are concerned, observed figures based on a large sample are available but this is not the case for typical process industry tasks.

Because so much judgement is involved, it is tempting for those who wish to do so to try to 'jiggle' the figures to get the answer they want. (A witness at an inquiry said, '... I tried to talk my staff into doing it as they, at least, work for me. And if they put in a number I didn't like, I could jiggle it around.'[14] He was not speaking particularly of figures on human error but there is no branch of hazard analysis to which his remarks are more applicable.) Anyone who uses estimates of human reliability outside the usual ranges (see later) should be expected to justify them.

Third, the actual error rates depend greatly on the degree of stress and distraction. The figures quoted are for typical process industry conditions but too much or too little stress can increase error rates (see Section 4.2.3, page 83).

Fourth, everyone is different and no-one can estimate the probability that he or she will make an error. All we can estimate is the probability that, if a large number of people were in a similar situation, errors would be made.

7.1 Why do we need to know human error rates?

In brief, because men are part of the total protective system and we cannot estimate the reliability of the total protective system unless we know the reliability of each part of it, including the human.

Consider the situation shown in Figure 7.1. When the alarm sounds and flashes in the control room the operator is expected to go outside, select the correct valve out of many and close it within, say, 10 minutes. We can estimate fairly accurately the reliability of the alarm and if we think it is too low it is easy to improve it. We can estimate roughly the reliability of the valve — the probability that the operator will be able to turn the handle and that the flow will actually stop — and if we think it is too low we can use a better quality valve or two in series. But what about the man in the middle? Will he always do what he is expected to do?

In my time in industry, people's opinions varied from one extreme to another. At times people, particularly design engineers, said that it was reasonable to expect him to always close the right valve. If he did not he should be reprimanded. I hope that the incidents described in Chapter 2 will have persuaded readers of the impracticality of this approach.

At other times people, particularly managers responsible for production, have said that it is well-known that men are unreliable. We should therefore install fully automatic equipment, so that the valve is closed automatically when the alarm condition is reached.

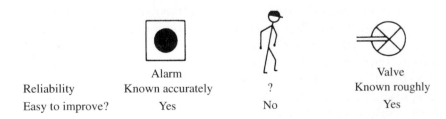

	Alarm		Valve
Reliability	Known accurately	?	Known roughly
Easy to improve?	Yes	No	Yes

Figure 7.1 Reliabilities in a man/machine system

Both these attitudes are unscientific. We should not say, 'The operator always will ... ' or 'The operator never will ... ' but ask, 'What is the probability that the operator will ... '. Having agreed a figure we can feed it into our calculations. If the consequent failure rate is too high, we can consider a fully automatic system. If it is acceptable, we can continue with the present system.

Operators usually notice that a plant is approaching a dangerous condition and take action before the emergency trip operates. If they did not, we would have to install more reliable trip systems than we normally install. Reliability engineers often assume that operators will detect an approach to trip conditions four times out of five or nine times out of ten and that only on the fifth or tenth occasion will the trip have to operate.

An example may make this clearer. Suppose the temperature in a vessel is automatically controlled by varying the heat input. In addition, an independent trip isolates the source of heat if the temperature in the tank rises above the set point. The trip fails once in 2 years (a typical figure) and is tested every 4 weeks (another typical figure). The trip fails at random so on average it will be dead for 2 weeks every 2 years or for 2% of the time. Its fractional dead time or probability of failure on demand is 0.02.

Assume the temperature rises above the set point 5 times/year and a hazardous situation would develop unless the operator or the trip intervenes. (This is not a typical figure; every situation is different.) Assume also that the operator spots the rising level and intervenes on 4 out of the 5 occasions. The demand rate on the trip is once/year, the probability that it will be in a failed state is 0.02, and the vessel will overflow once every 50 years. The consequences are such that this is considered tolerable. But how confident are we that the operator will intervene 4 times out of 5? Even if experience shows that he did so when the plant was new, will he continue to do so when manning is reduced and the operator has extra duties? Will anyone even ask this question?

If you are a designer, control engineer or plant manager, do you know how often you are expecting your operators to intervene before your trips operate? Is the answer realistic?

This simple example shows that installing a trip has not removed our dependence on the operator. In addition, we are also dependent on those who design, construct, install, test and maintain the trip. Automation does not remove dependence on people; it merely transfers some or all of the dependence onto different people.

7.2 Human error rates – a simple example

The following are some figures I have used for failure rates in the situation described — that is, the probability that a typical operator will fail to close the right valve in the required time, say 10 minutes. The estimates are conservative figures based on the judgements of experienced managers.

	Failure probability
(1) *When failure to act correctly will result in a serious accident such as a fire or explosion.* The operator's failure rate will not really be as high as this, but it will be very high (more than 1 in 2) as when someone is in danger, or thinks he is in danger, errors increase[15]. We should assume 1 in 1 for design purposes.	1 (1 in 1)
(2) *In a busy control room.* This may seem high but before the operator can respond to the alarm another alarm may sound, the telephone may ring, a fitter may demand a permit-to-work and another operator may report a fault outside. Also, on many plants the control room operator cannot leave the control room and has to contact an outside man, by radio or loudspeaker, and ask him to close the valve. This provides opportunities for misunderstanding.	0.1 (1 in 10)
(3) *In a quiet control room.* Busy control rooms are the more common, but control rooms in storage areas are usually quieter places with less stress and distraction. In some cases a figure between 1 in 10 and 1 in 100 can be chosen.	0.01 (1 in 100)
(4) *If the valve is immediately below the alarm.* The operator's failure rate will be very low but an occasional error may still occur. A failure rate of 1 in 1000 is about the lowest that should be assumed for any process operation — for example, failure to open a valve during a routine operation, as described in Section 2.2 (page 13).	0.001 (1 in 1000)

7.3 A more complex example

Figure 7.2 shows the fault tree for loss of level in a distillation column followed by breakthrough of vapour at high pressure into the downstream storage tank. A line diagram and further details including a detailed quantitative hazard analysis are given in Reference 1.

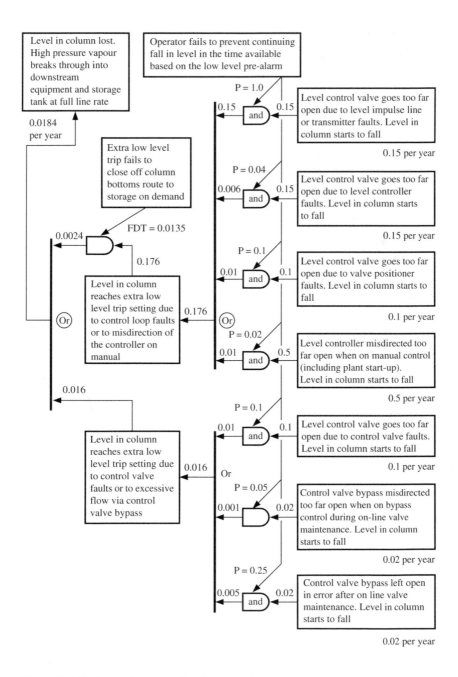

Figure 7.2 Fault tree for vapour breakthrough from high pressure distillation column into downstream equipment due to loss of level in distillation column base (original design)

133

The right-hand column of the tree shows the 'bottom events', the initiating events that can lead to loss of level. An alarm warns the operator that the level is falling. The tree shows the probabilities that the operator will fail to prevent a continuing fall in level in the time available.

For the top branch this probability, P, is 1 because the alarm and the level indicator in the control room would also be in a failed condition. The operator could not know that the level was falling.

For the other branches, values of P between 0.02 and 0.25 have been estimated. They include an allowance ($P = 0.005$) for the coincidental failure of the alarm.

For the fourth branch P has been assumed to be 0.02 (2% or 1 in 50) as the operator is normally in close attendance when a controller is on manual and correction is possible from the control room.

For the second branch P has been assumed to be 0.04 (4% or 1 in 25) as conditions are more unexpected and the operator is more likely to be busy elsewhere.

For the third and fifth branches $P = 0.1$ (10% or 1 in 10) as the operator has to ask an outside man to adjust a manual valve. The control room operator may delay making the request in order to make sure that the level really is falling.

For the sixth branch $P = 0.05$ (5% or 1 in 20). The outside operator has to be contacted but he should be near the valve and expecting to be contacted.

Finally, for the last branch $P = 0.25$ (25% or 1 in 4). The fault is unusual and the outside man may overlook it.

In the study of which Figure 7.2 formed a part, the estimates of operator reliability were agreed between an experienced hazard analyst and the commissioning manager. This reduces the chance that the figures will be 'jiggled' (see quotation on page 129) to get a result that the designer, or anyone else, would like. The figures used apply to the particular plant and problem and should not be applied indiscriminately to other problems. For example, if loudspeakers are inaudible in part of the plant or if several outside operators can respond to a request, so that each leaves it to the others (see Section 3.3, page 53), the probability that they will fail will be high.

In another similar study Lawley[2] used the figures set out in Table 7.1 in a form used by Lees[3].

Table 7.1 Industrial fault tree analysis: operator error estimates

Crystallizer plant	Probability
Operator fails to observe level indicator or take action	0.04
Operator fails to observe level alarm or take action	0.03
	Frequency events/y
Manual isolation valve wrongly closed (p)	0.05 and 0.1
Control valve fails open or misdirected open	0.5
Control valve fails shut or misdirected shut (l)	0.5

Propane pipeline	Time available	Probability
Operator fails to take action:		
• to isolate pipeline at planned shutdown		0.001
• to isolate pipeline at emergency shutdown		0.005
• opposite spurious tank blowdown given alarms and flare header signals	30 min	0.002
• opposite tank low level alarm		0.01
• opposite tank level high given alarm with	5–10 min	
(a) controller misdirected or bypassed when on manual		0.025
(b) level measurement failure		0.05
(c) level controller failure		0.05
(d) control valve or valve positioner failure		0.1
• opposite slowly developing blockage on heat exchanger revealed as heat transfer limitation		0.04
• opposite pipeline fluid low temperature given alarm	5 min	0.05
• opposite level loss in tank supplying heat transfer medium pump given no measurement (p)	5 min	0.2
• opposite tank blowdown without prior pipeline isolation given alarms which operator would not regard as significant and pipework icing	30 min	
(a) emergency blowdown		0.2
(b) planned blowdown		0.6
• opposite pipeline fluid temperature low given alarm	limited	0.4
• opposite backflow in pipeline given alarm	extremely short	0.8
• opposite temperature low at outlet of heat exchanger given failure of measuring instrument common to control loop and alarm		1

Continued overleaf

Table 7.1 (cont'd) Industrial fault tree analysis: operator error estimates

---

Misvalving in changeover of two-pump set (standby pump left valved open, working pump left valved in)	0.0025/ changeover
Pump in single or double pump operation stopped manually without isolating pipeline	0.01/ shutdown
LP steam supply failure by fracture, blockage or isolation error (p)	**Frequency** 0.1/y
Misdirection of controller when on manual (assumed small proportion of time)	1/y

---

Notes:
l = literature valve
p = plant value
Other values are assumptions

7.4 Other estimates of human error rates

Bello and Columbori[4] have devised a method known as TESEO (Technica Empirica Stima Errori Operati). The probability of error is assumed to be the product of five factors — K_1-K_5 — which are defined and given values in Table 7.2.

Table 7.3 (pages 138–139) is a widely quoted list of error probabilities taken from the US Atomic Energy Commission Reactor Safety Study (the Rasmussen Report)[6].

Swain[5] has developed a method known as THERP (Technique for Human Error Rate Prediction). The task is broken down into individual steps and the probability of error estimated for each, taking into account:

- the likelihood of detection;
- the probability of recovery;
- the consequence of the error (if uncorrected);
- a series of 'performance shaping factors' such as temperature, hours worked, complexity, availability of tools, fatigue, monotony, and group identification (about 70 in total).

Table 7.2 TESEO: error probability parameters

Type of activity

	K_1
Simple, routine	0.001
Requiring attention, routine	0.01
Not routine	0.1

Temporary stress factor for routine activities

Time available, s	K_2
2	10
10	1
20	0.5

Temporary stress factor for non-routine activities

Time available, s	K_2
3	10
30	1
45	0.3
60	0.1

Operator qualities

	K_3
Carefully selected, expert, well-trained	0.5
Average knowledge and training	1
Little knowledge, poorly trained	3

Activity anxiety factor

	K_4
Situation of grave emergency	3
Situation of potential emergency	2
Normal situation	1

Activity ergonomic factor

	K_5
Excellent microclimate, excellent interface with plant	0.1
Good microclimate, good interface with plant	1
Discrete microclimate, discrete interface with plant	3
Discrete microclimate, poor interface with plant	7
Worst microclimate, poor interface with plant	10

Table 7.3 General estimates of error probability used in US Atomic Energy Commission reactor safety study

Estimated error probability	Activity
10^{-4}	Selection of a key-operated switch rather than a non-key switch (this value does not include the error of decision where the operator misinterprets the situation and believes the key switch is the correct choice).
10^{-3}	Selection of a switch (or pair of switches) dissimilar in shape or location to the desired switch (or pair of switches), assuming no decision error. For example, operator actuates large-handled switch rather than small switch.
3×10^{-3}	General human error of commission, e.g. misreading label and therefore selecting wrong switch.
10^{-2}	General human error of omission where there is no display in the control room of the status of the item omitted, e.g. failure to return manually operated test valve to proper configuration after maintenance.
3×10^{-3}	Errors of omission, where the items being omitted are embedded in a procedure rather than at the end as above.
3×10^{-2}	Simple arithmetic errors with self-checking but without repeating the calculation by re-doing it on another piece of paper.
$1/x$	Given that an operator is reaching for an incorrect switch (or pair of switches), he selects a particular similar appearing switch (or pair of switches), where x = the number of incorrect switches (or pairs of switches) adjacent to the desired switch (or pair of switches). The $1/x$ applies up to 5 or 6 items. After that point the error rate would be lower because the operator would take more time to search. With up to 5 or 6 items he does not expect to be wrong and therefore is more likely to do less deliberate searching.
10^{-1}	Given that an operator is reaching for a wrong motor operated valve (MOV) switch (or pair of switches), he fails to note from the indicator lamps that the MOV(s) is (are) already in the desired state and merely changes the status of the MOV(s) without recognizing he had selected the wrong switch(es).

Continued opposite

Table 7.3 (cont'd) General estimates of error probability used in US Atomic Energy Commission reactor safety study

~ 1.0	Same as above, except that the state(s) of the incorrect switch(es) is (are) *not* the desired state.
~ 1.0	If an operator fails to operate correctly one of two closely coupled valves or switches in a procedural step, he also fails to operate correctly the other valve.
10^{-1}	Monitor or inspector fails to recognize initial error by operator. Note: with continuing feedback of the error on the annunciator panel, this high error rate would not apply.
10^{-1}	Personnel on different work shift fail to check condition of hardware unless required by check-list or written directive.
5×10^{-1}	Monitor fails to detect undesired position of valves, etc., during general walk-around inspections, assuming no check-list is used.
0.2–0.3	General error rate given very high stress levels where dangerous activities are occurring rapidly.
$2^{(n-1)}x$	Give severe time stress, as in trying to compensate for an error made in an emergency situation, the initial error rate, x, for an activity doubles for each attempt, n, after a previous incorrect attempt, until the limiting condition of an error rate of 1.0 is reached or until time runs out. This limiting condition corresponds to an individual's becoming completely disorganized or ineffective.
~ 1.0	Operator fails to act correctly in first 60 seconds after the onset of an extremely high stress condition, e.g. a large LOCA (loss of cooling accident).
9×10^{-1}	Operator fails to act correctly after the first 5 minutes after the onset of an extremely high stress condition.
10^{-1}	Operator fails to act correctly after the first 30 minutes in an extreme stress condition.
10^{-2}	Operator fails to act correctly after the first several hours in a high stress condition.
x	After 7 days after a large LOCA, there is a complete recovery to the normal error rate, x, for any task.

An advantage of THERP is its similarity to the methods used for estimating the reliability of equipment: each component of the equipment is considered separately and its basic failure rate is modified to allow for environmental conditions such as temperature and vibration. The method thus appeals to reliability engineers. The disadvantages of THERP are the amount of time and expert effort required. Many human factors experts believe that any gain in accuracy compared with the simpler methods already described is out of all proportion to the effort expended.

A study of batch chemical operations produced the following results:

	Probability per batch
Ingredient omitted	2.3×10^{-4} (1 in 4375 batches)
Ingredient undercharged	1.1×10^{-4} (1 in 8750 batches)
Ingredient overcharged	2.6×10^{-4} (1 in 3900 batches)
Wrong ingredient added	2.3×10^{-4} (1 in 4375 batches)
Total errors	8.3×10^{-4} (1 in 1200 batches)

These error rates seem rather low. However they do not include:

- errors which were not reported — estimated at 50% of the total;
- errors which were corrected — estimated at 80% of the total;
- errors which were too small to matter — estimated at 80% of the total.

If these are included, the total error rate becomes 4×10^{-2} per batch or 1 in 25 batches.

There are on average four charging operations per batch so the error rate becomes 10^{-2} per operation (1 in 100 operations) which is in line with other estimates.

Embrey[23] gives more detailed descriptions of these and other techniques, including a description of the Influence Diagram Approach, a technique for estimating the influence of training and instructions on the probability of error.

The University of Birmingham has developed a database of human error probabilities called CORE-DATA (Computerised Operator Reliability and Error DATAbase)[24].

Taylor-Adams and Kirwan[25] have published some human error failure rates based on one company's experience but their paper does not show whether the errors were due to slips or lapses of attention, poor training or instructions or non-compliance with instructions. Their data could be used to estimate future error rates in the company of origin, if there is no evidence of change, but not in other companies. Data on slips and lapses of attention are generic; data on errors due to poor training or instructions or non-compliance with instructions apply only to the place where they were measured (see Section 7.10, page 149).

7.5 Two more simple examples

7.5.1 Starting a spare pump

As an example of the uses of some of these figures, let us consider another simple process operation: starting up a spare pump after the running pump has tripped out.

Many analysts would use the simple approach of Section 7.2 and assume that the job will be done correctly 99 times out of 100 in a normal, low stress situation, and rather less often — perhaps 9 times out of 10 — if the stress is high — say the operator knows that the plant will shut down in 5 minutes if the spare pump is not started up correctly.

Let us see if a more analytical approach is helpful. The task can be split into a number of steps:

(1) Walk to pump.
(2) Close delivery valve of failed pump.
(3) Close suction valve of failed pump.
(4) Open suction valve of spare pump.
(5) Press start button.
(6) Open delivery valve of spare pump.

It does not matter if the operator forgets to carry out steps (2) and (3), so there are four steps which have to be carried out correctly.

Step (1) is included as perhaps one of the commonest sources of error is failing to carry out this step — that is, the operator forgets the whole job because he has other things on his mind, or goes to the wrong pump.

From Table 7.3, the lines which seem most applicable, in a low stress situation, are lines three and five.

Estimated error probability	Activity
3×10^{-3}	General human error of commission — for example, misreading label and therefore selecting wrong switches
3×10^{-3}	Errors of omission, where the items being omitted are embedded in a procedure

There are four critical steps, so the total probability of error is:

$12 \times 10^{-3} = 0.012$ (1 in 80)

Table 7.3 is not much help in a condition of moderate stress, though it does consider very high stress situations (last five lines), so try applying Table 7.2.

Type of activity:	Requiring attention, routine	$K_1 = 0.01$
Stress factor:	More than 20 secs available	$K_2 = 0.5$
Operator qualities:	Average knowledge and training	$K_3 = 1$
Activity anxiety factor:	Potential emergency	$K_4 = 2$
Activity ergonomic factor:	Good microclimate, good interface with plant	$K_5 = 1$

Probability of error $= K_1\, K_2\, K_3\, K_4\, K_5 = 0.01$

If these figures are assumed to apply to each step, the total probability of error is 0.04 (1 in 25).

However if we assume each step is 'simple', rather than one 'requiring attention', the error rate is 10 times lower for each step, and is now 1 in 250 for the task as a whole. This illustrates the limitation of these techniques, when successive figures in a table differ by an order of magnitude, and shows how easy it is for an unscrupulous person to 'jiggle' the figures to get the answer he or she wants. The techniques are perhaps most valuable in estimating relative error probabilities rather than absolute values.

It is interesting to compare these estimates with the reliability of an automatic start mechanism. A typical failure rate will be 0.25/year and if the mechanism is tested every 4 weeks, its fractional dead time (probability of failure on demand) will be:

$\frac{1}{2} \times 0.25 \times 4/52 = 0.01$ or 1 in 100

None of our estimates of human reliability is lower than this, thus confirming the instinctive feel of most engineers that in this situation an automatic system is more reliable and should be installed if failure to start the pump has serious consequences, but it is hardly justified in the normal situation.

142

7.5.2 Filling a tank

Suppose a tank is filled once/day and the operator watches the level and closes a valve when it is full. The operation is a very simple one, with little to distract the operator who is out on the plant giving the job his full attention. Most analysts would estimate a failure rate of 1 in 1000 occasions or about once in 3 years. In practice, men have been known to operate such systems for 5 years without incident. This is confirmed by Table 7.2 which gives:

$K_1 = 0.001$
$K_2 = 0.5$
$K_3 = 1$
$K_4 = 1$
$K_5 = 1$

Failure rate $= 0.5 \times 10^{-3}$ or 1 in 2000 occasions (6 years).

An automatic system would have a failure rate of about 0.5/year and as it is used every day testing is irrelevant and the hazard rate (the rate at which the tank is overfilled) is the same as the failure rate, about once every 2 years. The automatic equipment is therefore less reliable than an operator.

Replacing the operator by automatic equipment will increase the number of spillages unless we duplicate the automatic equipment or use Rolls Royce quality equipment (if available). (For a more detailed treatment of this problem see Reference 7.)

As mentioned in Section 7.1 (page 130), if we replace an operator by automatic equipment we do not, as is often thought, eliminate the human element. We may remove our dependence on the operator but we are now dependent on the people who design, construct, install, test and maintain the automatic equipment. We are merely dependent on different people. It may be right to make the change (it was in one of the cases considered, not the other) as these people usually work under conditions of lower stress than the operator, but do not let us fool ourselves that we have removed our dependence on people.

7.5.3 More opportunities – more errors

In considering errors such as those made in starting a pump, filling a tank, etc., do not forget that the actual *number* of errors made by an operator, as distinct from the *probability* of errors, depends on the number of times he is expected to start a pump, fill a tank, etc.

I once worked in a works that consisted of two sections: large continuous plants and small batch plants. The foremen and operators on the batch plants had a poor reputation as a gang of incompetents who were always making mistakes: overfilling tanks, putting material in the wrong tank, etc. Some of the best men on the continuous plants were transferred to the batch plants but with little effect. Errors *rates* on the batch plants were actually lower than on the continuous plants but there were more opportunities for error; pumps were started up, tanks filled, etc., many more times per day.

7.5.4 The effect of the time available

The more time we have to take action, the less likely we are to take the wrong action and attempts have been made to quantify the effect (see Table 7.3, last five items). Skill-based actions, discussed in Chapter 2, are automatic or semi-automatic and need less time for successful completion than rule-based actions which in turn need less time than knowledge-based actions (both discussed in Chapter 3.) Of course, if the action required is beyond the ability of the person concerned or he decides not to carry out the action, extra time will make no difference.

Plant designers sometimes assume that if someone has 30 minutes or more in which to act, the probability that he will fail to do so is small and can be ignored. However, this assumes that the action required is within the person's physical or mental ability (Chapter 4), that he does not, for any reason, decide to ignore instructions (Chapter 5), that he has been told to carry out the action (if it is rule-based) and that he has been given the necessary training (if it is knowledge-based) (see Chapter 3). These assumptions are not always true.

7.6 Button pressing

The American Institute for Research has published a series of papers on the reliability of simple operations such as those used in operating electronic equipment[8]. The application and limitations of the data can be illustrated by applying it to one of the button-pressing operations described in Section 2.4 (page 25), the operation of a beverage vending machine.

The pushbuttons are considered under several headings:

The first is size. The probability that the correct button will be pressed depends on the size as shown below:

Miniature	0.9995
$\frac{1}{2}$ inch	0.9999
more than $\frac{1}{2}$ inch	0.9999 ←

The pushbuttons on the beverage machines are $1\frac{1}{2}$ inch by $\frac{5}{8}$ inch so I have put an arrow against the last item.

Next we consider the number of pushbuttons in the group. Single column or row:

1–5	0.9997
4–10	0.9995 ←
11–25	0.9990

The next item to be considered is the number of pushbuttons to be pushed:

2	0.9995
4	0.9991
8	0.9965

On a beverage machine only one button has to be pushed so I assumed that the probability of success is 0.9998.

The fourth item is the distance between the edges of the buttons:

$\frac{1}{8}$ inch – $\frac{1}{4}$ inch	0.9985 ←
$\frac{3}{8}$ inch – $\frac{1}{2}$ inch	0.9993
$\frac{1}{2}$ inch up	0.9998

The fifth item is whether or not the button stays down when pressed:

Yes	0.9998 ←
No	0.9993

The final item is clarity of the labelling:

At least two indications of control positioning	0.9999
Single, but clear and concise indication of control positioning	0.9996 ←
A potentially ambiguous indication of control positioning	0.9991

The total probability of success is obtained by multiplying these six figures together:

Reliability = 0.9999 × 0.9995 × 0.9998 × 0.9985 × 0.9998 × 0.9996 = 0.9971
(that is, three errors can be expected in every 1000 operations)

I actually got the wrong drink about once in 50 times (that is, 20 in 1000 times) that I used the machines. Perhaps I am seven times more careless than the average person, or inadequately trained or physically or mentally incapable. It is more likely that the difference is due to the fact that the method of calculation makes no allowance for stress or distraction (ignored in this mechanistic approach) and that the small amount present is sufficient to increase my error rate seven times. (The machines are in the corridor, so I may talk to the people who pass, or I may be thinking about the work I have just left.)

Reference 31 lists the factors which affect the probability of an error in keying a telephone number.

7.7 Non-process operations

As already stated, for many assembly line and similar operations error rates are available based not on judgement but on a large data base. They refer to normal, not high stress, situations. Some examples follow. Remember that many errors can be corrected and that not all errors matter (or cause degradation of mission fulfilment, to use the jargon used by many workers in this field).

A.D. Swain[9] quotes the following figures for operations involved in the assembly of electronic equipment:

	Error rate per operation
Excess solder	0.0005
Insufficient solder	0.002
Two wires reversed	0.006
Capacitor wired backwards	0.001

He quotes the following error rates for inspectors measuring the sizes of holes in a metal plate:

Errors in addition, subtraction and division	0.013
Algebraic sign errors	0.025
Measurement reading errors	0.008
Copying errors per six-digit number	0.004

These should not, of course, be used uncritically in other contexts.

7.8 Train driver errors

Section 2.9.3 (page 38) discusses errors by train drivers who passed signals at danger, so it may be interesting to quote an estimate of their error rates[10].

In 1972, in the Eastern Region of British Rail, 91 signals were passed at danger (information from British Rail). 80–90% of these, say 83, were due to drivers' errors. The rest were due to other causes such as signals being suddenly altered as the train approached. It was estimated that many incidents are not reported at the time and that the true total was 2–2.5 times this figure, say 185. The total mileage covered in the Eastern Region in 1972 was 72.2×10^6 miles (information from British Rail). The spacing of signals is about 1 every 1200 yards (0.7 mile) on the main lines; rather less frequent on branch lines. Say 0.75 mile on average. Therefore, in 1972, $72.2 \times 10^6/0.75 = 96 \times 10^6$ signals were passed. Therefore, the chance that a signal will be passed at danger is 1 in $96 \times 10^6/185 = 1$ in 5×10^5 signals approached (*not* signals at danger approached). If we assume that between 1 in 10 and 1 in 100 (say 1 in 35) signals approached is at danger then a signal will be passed at danger once every $5 \times 10^5/35 \approx 10^4$ occasions that a signal at danger is approached. This is at the upper end of the range quoted for human reliability[26].

185 signals passed at danger in 72.2 million train-miles is equivalent to 2.6 incidents per million train-miles. Taylor and Lucas[27] quote 2 incidents per million train-miles for 1980 and 3 for 1988, in good agreement. However, nearly half their incidents were due to insufficient braking so that the train came to a halt a short distance past the signal. Such incidents may be excluded from the figures I was given. Even so, Taylor and Lucas' figures suggest that my estimates are not too far out.

Another recent estimate is once in 17,000 signals at danger approached[30].

7.9 Some pitfalls in using data on human reliability

7.9.1 Checking may not increase reliability

If a man knows he is being checked, he works less reliably. If the error rate of a single operator is 1 in 100, the error rate of an operator plus a checker is certainly greater than 1 in 10,000 and may even be greater than 1 in 100 — that is, the addition of the checker may actually increase the overall error rate.

A second man in the cab of a railway engine is often justified on the grounds that he will spot the driver's errors. In practice, the junior man is usually reluctant to question the actions of the senior[16]. In the same way, the second officer on the flight deck of an aircraft may be reluctant to question the actions of the captain. One reason suggested for Australia's outstandingly good air safety record is that they have a culture in which second officers are not reluctant to question the actions of the captain[17].

If two men can swop jobs and repeat an operation then error rates come down. Reference 11 describes the calibration of an instrument in which one man writes down the figures on a check-list while the other man calls them out. The two men then change over and repeat the calibration. The probability of error was put at 10^{-5}.

Requiring a man to sign a statement that he has completed a task produces very little increase in reliability as it soon becomes a perfunctory activity[11] (see Section 6.5, page 118).

The following incident shows how checking procedures can easily lapse.

A man was asked to lock off the power supply to an electric motor and another man was asked to check that he had done so. The wrong breaker was shown on a sketch given to the men but the right one was shown on the permit and the tag and was described at a briefing. The wrong motor was locked off. The report on the incident states, 'In violation of the site conduct-of-operations manual, which requires the acts of isolation and verification to be separated in time and space, the isolator and the verifier worked together to apply the lockout. Both isolator and the verifier used the drawing with the incorrectly circled breaker to identify the breaker to be locked out; neither referred to the lockout/tagout permit or the lockout tags, although both initialled them.'[28]

Some organizations expect the man who is going to carry out the maintenance to check the isolations. Other organizations encourage him to do so but do not insist on it, as they believe this will weaken the responsibility of the isolator. Either way, the man who is going to carry out the maintenance will be well motivated. He is the one at risk.

7.9.2 Increasing the number of alarms does not increase reliability proportionately

Suppose an operator ignores an alarm on 1 in 100 of the occasions on which it sounds. Installing another alarm (at a slightly different setting or on a different parameter) will not reduce the failure rate to 1 in 10,000. If the operator is in a state in which he ignores the first alarm, then there is a more than average chance that he will ignore the second. (In one plant there were five alarms in series. The designer assumed that the operator would ignore each alarm on one occasion in ten, the whole lot on one occasion in 100,000!).

7.9.3 If an operator ignores a reading he may ignore the alarm

Suppose an operator fails to notice a high reading on 1 occasion in 100 — it is an important reading and he has been trained to pay attention to it.

Suppose that he ignores the alarm on 1 occasion in 100. Then we cannot assume that he will ignore the reading and the alarm on one occasion in 10,000. On the occasion on which he ignores the reading the chance that he will ignore the alarm is greater than average.

7.9.4 Unusually high reliability

I have suggested above that about 1 in 1000 actions is the lowest error rate that should be assumed for any process operation and Table 7.3 and Section 7.6 give 1 in 10,000 as the error rate for the simplest tasks, such as pressing a button. According to Rasmussen and Batstone[18], in certain circumstances people are capable of much higher reliability. They quote a net error rate (after recovery) of 1 in 10^7 per opportunity for air traffic controllers and for flight deck operations on aircraft carriers. They suggest that to achieve these high reliabilities the operator:

- must be able to recover from the effects of error;
- must be thoroughly familiar with the task;
- must be in continuous interaction with a rapidly changing system where errors, if not corrected, will lead to loss of control. In process operations the operator only interacts with the system from time to time; he makes an adjustment and then turns his attention elsewhere. Also, process operations are usually designed so that there is a large margin between normal operation and loss of control and the operators are not continuously operating at the boundaries of safe operation.

While the first edition of this book was in the press a review appeared of attempts to validate human reliability assessment techniques[19]. The author concludes, 'Even the most comprehensive study of assessment techniques cannot tell us whether any of the methods it evaluated are worth much further consideration. Perhaps the only real message to emerge is that Absolute Probability Judgment (APJ) was the least worse of the methods evaluated.'

APJ is another term for experienced judgement, gut feel or, if you prefer, guessing.

7.10 Data on equipment may be data on people

Does data on the reliability of equipment tell us something about the equipment or does it tell us more about the people who maintain and operate the equipment?

Consider light bulbs as an example of a mass-produced article usually used under standard conditions. We do not expect the failure rate to vary from one user to another and we are confident that failure data will tell us something about the inherent properties of the bulbs. We would use someone else's data with confidence. Even here, however, if someone reported an unusually high failure rate we would question the conditions of use (were they used horizontally or upside down or subject to vibration?) or wonder if the user (or supplier) had treated them roughly.

Instruments come into the same category as light bulbs. The failure rate gives us information about the equipment and does not vary by more than a factor of about four even in a harsh environment[20].

Mechanical equipment is different; the failure rate depends on the conditions of use and the maintenance policy. These vary greatly from one location to another and great caution is needed in using other people's data.

Beverage vending machines are a simple example. Only rarely do the machines in the office fail to deliver (or deliver cold). If you have tried to use similar machines in public places you will know that they often fail. The design may be the same but the customers treat them differently and there is probably no system for reporting and repairing damage.

A more technical example is bellows. A study in one company showed that 1 in 50 failed per year and had to be replaced. The main cause of failure was poor installation and the second most common cause was wrong specification of the material of construction. The failure rate varied a great deal between one factory and another and told us more about the training and motivation of the maintenance and construction teams and the quality of the supervision and inspection than it told us about the bellows. The information was, nevertheless, still useful. It told us that bellows cannot withstand poor installation and are therefore best avoided when hazardous materials are handled. Expansion loops should be used instead. If bellows are used we should specify the method of installation in detail, train the maintenance team and check that the bellows are properly installed.

The other cause of bellows failure — incorrect specification — is easy to put right once it is recognized. The failure rate due to this cause tells us about the knowledge of the specifiers and the company's procedures for transferring information from one design section to another.

Similarly, data on pipework failures tell us about the quality of design and construction rather than the quality of the pipes used (Section 9.1, page 168).

Machinery sometimes fails because it has not been lubricated regularly. The failure data measure the training or competence of the operating team and tell us nothing about the machinery[21].

A man had three Ford cars and crashed each of them so he decided to try another make. Does this tell us something about Ford cars or about the man?

7.11 Who makes the errors?

Designers, maintenance workers, operators and others all make errors. In some activities maintenance errors may predominate, in others operating errors and so on. Scott and Gallaher[22] quote the following for the distribution of errors leading to valve failure in pressurized water nuclear reactors:

Physical causes	54%
Human causes	46
Administrative errors	7%
Design errors	8
Fabrication errors	4
Installation errors	5
Maintenance errors	16
Operator errors	7

Chapters 8–11 consider in turn errors made during design, construction, maintenance and operations, regardless of the type of error involved.

7.12 Conclusions

Enthusiasts for the quantification of risk sometimes become so engrossed in the techniques that they forget the purpose and limitations of the study, so it may be worth restating them. The purpose of risk quantification is to help us decide whether or not the risk is tolerable. There are risks so high that we do not tolerate them, risks so small that we accept them and in between we reduce them if the costs of doing so are not excessive. In the UK this philosophy has been clearly spelt out by the Health and Safety Executive[29]. In some other countries employers are, in theory, expected to remove all hazards, however unlikely to occur or small in their consequences, but in practice this is impossible.

Risks are sometimes said to be due either to human failures or equipment failures. The latter are due to failures by designers or other people (or to a decision to accept an occasional failure) but we find it convenient to use in our calculations the resultant figures for the failure rate of the equipment rather than the failure rate of the designer.

If we decide (quantitatively or otherwise) that a risk is too high, the first step should be to see if we can use an inherently safer design (see Section

151

6.7.1, page 120). For example, if a flammable solvent is used, could a non-flammable one be used instead? If not, can the probability of a leak and fire be reduced by using better designs of equipment or, if that is not practicable, better procedures? Fault tree analysis, as in Figure 7.2, is widely used for estimating hazard rates and risks to life. *Its great virtue, however, is that it shows very clearly which failures, of equipment or people, contribute most to the probability of failure. It thus tells us where we should seek improvements.*

References in Chapter 7

1. Kletz, T.A. and Lawley, H.G., 1982, Chapter 2.1 of Green, A.E. (editor), *High Risk Safety Technology* (Wiley, Chichester, UK).
2. Lawley, H.G., 1980, *Reliability Engineering*, 1: 89.
3. Lees, F.P., 1983, The assessment of human reliability in process control, *Conference on Human Reliability in the Process Control Centre, Manchester* (North West Branch, Institution of Chemical Engineers).
4. Bello, G.C. and Columbori V., 1980, *Reliability Engineering*, 1(1): 3.
5. Swain, A.D. and Gutterman, H.E., 1980, *Handbook of Human Reliability Analysis with Emphasis on Nuclear Power Plant Applications, Report No. NUREG/CR–1278* (Sandia Laboratories, Albuquerque, New Mexico).
6. Atomic Energy Commission, 1975, *Reactor Safety Study — An Assessment of Accident Risk in US Commercial Nuclear Power Plant, Report No. WASH 1400* (Atomic Energy Commission, Washington, DC, USA).
7. Kletz, T.A., 1999, *Hazop and Hazan — Identifying and Assessing Process Industry Hazards*, 4th edition, Section 3.6.6 (Institution of Chemical Engineers, Rugby, UK).
8. Payne, D. *et al.*, 1964, *An Index of Electronic Equipment Operability, Report No.s AD 607161–5* (American Institute for Research, Pittsburgh, PA, USA).
9. Swain, A.D., 1973, *Seminar on Human Reliability* (UK Atomic Energy Authority, Risley, Warrington, UK).
10. Kletz, T.A., 1976, *Journal of Occupational Accidents*, 1(1): 95.
11. As Reference 6, Appendix III.
12. Gonner, E.C.K., 1913, *Royal Statistical Society Journal*. Quoted by Boreham, J. and Holmes, C., 1974, *Vital Statistics* (BBC Publications, London, UK).
13. Pitblado, R.M., Williams, J.C. and Slater, D.H., 1990, *Plant/Operations Progress*, 9(37): 169.
14. Akerm, W., quoted by Kunreither, H.C. and Linnerooth, J., 1983, *Risk Analysis and Decision Processes* (Springer-Verlag, Germany).
15. Livingstone-Booth, A. *et al.*, 1985, *Reliability Engineering*, 13(4): 211.
16. See, for example, *Railroad Accident Report: Head-on Collision of Two Penn Central Transportation Company Freight Trains near Pettisville, Ohio, Report*

No. NTSB–RAR–76–10, 1976 (National Transportation Safety Board, Washington, DC, USA).

17. Ramsden, J.M., *Flight International*, 1 December 1984, page 1449 and 26 January 1985, page 29.

18. Rasmussen, J. and Batstone, R., 1989, *Why Do Complex Organisational Systems Fail?*, page 35 (World Bank, Washington, DC, USA).

19. Williams, J.C., 1985, *Reliability Engineering*, 11(3): 149.

20. Lees, F.P., 1996, *Loss Prevention in the Process Industries*, 2nd edition, Section 13.6 (Butterworth-Heinemann, Oxford, UK).

21. Kletz, T.A., 1985, *Reliability Engineering*, 11: 185.

22. Scott, R.L. and Gallaher, R.B., *Operating Experience with Valves in Lightwater-reactor Nuclear Power Plants for the Period 1965–1978, Report No. NUREG/CR 0848* (quoted in *The MANAGER Technique*, November 1988 (Technica, London, UK)).

23. Embrey, D., 1994, *Guidelines for Preventing Human Error in Process Safety*, Chapter 5 (American Institute of Chemical Engineers, New York, USA).

24. Gibson, W.H. and Megaw, T.D., 1998, *Contract Research Report 245/1999, The Implementation of CORE-DATA, a Computerised Human Error Probability Database* (HSE Books, Sudbury, UK).

25. Taylor-Adams, S. and Kirwan, B., 1997, *Disaster Prevention and Management*, 6(5): 318.

26. Boyes, J.M.. and Semmens, P.W.B., 1972, private communication.

27. Taylor, R.K. and Lucas, D.A., in van der Schaaf, T.W., Lucas, D.A. and Hale, A.R. (editors), 1991, *Near Miss Reporting as a Safety Tool* (Butterworth-Heinemann, Oxford, UK).

28. *Operating Experience Weekly Summary*, 1999, No. 99–24, page 11 (Office of Nuclear and Facility Safety, US Department of Energy, Washington, DC, USA).

29. Health and Safety Executive, 1992, *The Tolerability of Risk from Nuclear Power Stations*, 2nd edition (HSE Books, Sudbury, UK). See also HSE Discussion Document, *Reducing Risks, Protecting People*, 1999 (HSE Books, Sudbury, UK).

30. Ford, R., *Modern Railways*, December 2000, page 45.

31. Gibson, W.H. and Megaw, T.D., 1999, *Loss Prevention Bulletin*, No. 146, page 13.

Some accidents that could be prevented by better design

8

'Don't take square-peg humans and try and hammer them into round holes. Reshape the holes as squares.'
P. Foley[12]

'It is much simpler to find a few good designers and inspectors than to staff, with assurance, potentially thousands of plants with skilled operators.'
C.W. Forsberg[13]

So far, in describing accidents due to human error, I have classified them by the type of error. This chapter describes some more accidents due to various sorts of human error that could be prevented by better design — in particular, by changing the design so as to remove opportunities for error.

As stated in Chapter 1, safety by design should always be our aim but often there is no reasonably practicable or economic way of improving the design and we have to rely on improvements to procedures. We cannot buy our way out of every problem. However, this chapter gives examples of cases where changes in design are practicable, at least on new plants, often without extra cost.

I do not, of course, wish to imply that accidents are due to the negligence of designers. Designers make errors for all the reasons that other people make errors: a lapse of attention, ignorance, lack of ability, a deliberate decision not to follow a code. Unlike operators they usually have time to check their work, so slips and lapses of attention are frequently detected before the design is complete. Just as we try to prevent some accidents by changing the work situation, so we should try to prevent other accidents by changing the design situation — that is, we should try to find ways of changing the design process so as to produce better designs. The changes necessary will be clearer when we have looked at some examples, but the main points that come out are:

- Cover important safety points in standards or codes of practice.
- Make designers aware of the reasons for these safety points by telling them about accidents that have occurred because they were ignored. As with

154

operating staff, discussion is better than writing or lecturing (see Section 3.3.3, page 54).
- Carry out hazard and operability studies[1] on the designs. As well as the normal Hazop on the line diagrams, an earlier Hazop on the flowsheet (or another earlier study of the design concept) may allow designers to avoid hazards by a change in design instead of controlling them by adding on protective equipment[2] (see Section 8.7, page 162).
- Make designers more aware of the concepts described in Section 8.7.

8.1 Isolation of protective equipment

An official report[3] described a boiler explosion which killed two men. The boiler exploded because the water level was lost. The boiler was fitted with two sight glasses, two low level alarms, a low level trip which should have switched on the water feed pump and another low level trip, set at a lower level, which should have isolated the fuel supply. All this protective equipment had been isolated by closing two valves. The report recommended that it should not be possible to isolate all the protective equipment so easily.

We do not know why the valves were closed. Perhaps they had been closed for maintenance and someone forgot to open them. Perhaps they were closed in error. Perhaps the operator was not properly trained. Perhaps he deliberately isolated them to make operation easier, or because he suspected the protective equipment might be out of order. It does not matter. It should not be possible to isolate safety equipment so easily. It is necessary to isolate safety equipment from time to time but each piece of equipment should have its own isolation valves, so that only the minimum number need be isolated. Trips and alarms should be isolated only after authorization in writing by a competent person and this isolation should be signalled in a clear way — for example, by a light on the panel, so that everyone knows that it is isolated.

In addition, although not a design matter, regular checks and audits of protective systems should be carried out to make sure that they are not isolated. Such surveys, in companies where they are not a regular feature, can bring appalling evidence to light. For example, one audit of 14 photo-electric guards showed that all had been bypassed by modifications to the electronics, modifications which could have been made only by an expert[4].

8.2 Better information display

The tyres on a company vehicle were inflated to a gauge pressure of 40 psi (2.7 bar) instead of the recommended 25 psi (1.7 bar). A tyre burst and the vehicle skidded into a hut. Again we do not know why the tyres were overinflated. It may have been due to ignorance by the man concerned, or possibly he was distracted while inflating them. However, the factory management found a simple way of making such errors less likely — they painted the recommended tyre pressures above the wheels of all their vehicles.

There is another point of interest in this incident. The company concerned operated many factories, but only the one where the incident occurred made the change. The other factories were informed, but decided to take no action — or never got round to it. This is a common failing. When an accident has occurred to us we are willing to make a change to try to prevent it happening again, so willing that we sometimes over-react. When the accident has happened elsewhere, the safety adviser has to work much harder to persuade people to make a change. A small fire at our place of work has more effect than a large fire elsewhere in the country or a conflagration overseas.

Another example of prevention of errors by better information display is shown in Figure 8.1[5].

8.3 Pipe failures

About half of the large leaks that occur on oil and chemical plants are due to pipe failures. These have many causes. Here are accounts of a few that could have been prevented by better design[6].

8.3.1 Remove opportunities for operator errors

The wrong valve was opened and liquid nitrogen entered a mild steel line causing it to disintegrate. This incident is similar to those described in Chapter 2. For one reason or another an error is liable to be made sooner or later and an engineering solution is desirable. The pipe could be made of stainless steel, so that it would not matter if the valve was opened, or the valve could be an automatic one, kept shut by a low temperature trip. Designers would never design a plant so that operation of a valve in error caused equipment to be overpressured; they would install a relief valve or similar device. They are less willing to prevent equipment getting too cold (or too hot). Design codes and procedures should cover this point[7].

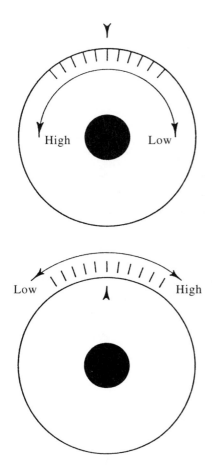

Figure 8.1 Puzzle — which way do you turn the knob to increase the reading? It is better to put the scale on the base-plate instead of the knob. There is then no doubt which way the knob should be turned.

8.3.2 Remove opportunities for construction errors

A fire at a refinery was caused by corrosion of an oil pipeline just after the point at which water had been injected[8] (Figure 8.2(a), page 158). A better design is shown in Figure 8.2(b) (page 158). The water is added to the centre of the oil stream through a nozzle so that it is immediately dispersed. However a plant that decided to use this system found that corrosion got worse instead of better. The nozzle had been installed pointing upstream instead of downstream (Figure 8.2(c), page 158).

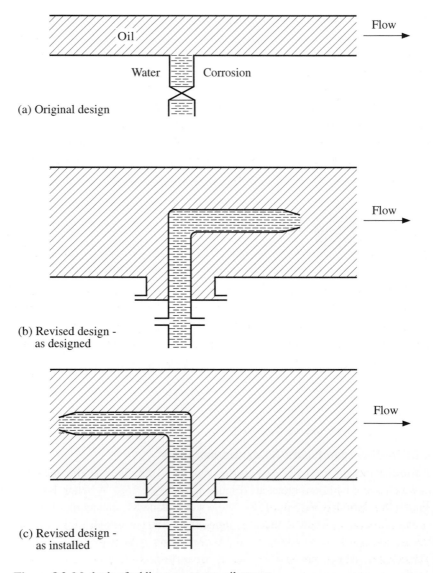

(a) Original design

(b) Revised design -
    as designed

(c) Revised design -
    as installed

Figure 8.2 Methods of adding water to an oil stream

Once the nozzle was installed it was impossible to see which way it was pointing. On a better design there would be an external indication of the direction of flow, such as an arrow, or, better still, it would be impossible to assemble the equipment incorrectly. We should design, when possible, so as to remove the opportunity for construction and maintenance errors as well as operating errors.

8.3.3 Design for all foreseeable conditions

A bellows was designed for normal operating conditions. When it was steamed through at a shutdown, it was noticed that one end was 7 inches higher than the other, although it was designed for a maximum deflection of ± 3 inches. During normal operation the maximum deflection was 1 inch.

The man who carried out the detailed design of the pipework was dependent on the information he received from other design sections, particularly the process engineering section. The design organization should ensure that he receives information on transient conditions, such as start-up, shutdown and catalyst regeneration, as well as normal operating conditions.

8.4 Vessel failures

These are very rare, much less common than pipework failures; some do occur although only a few of these could be prevented by better design. Many are the result of treating the vessel in ways not foreseen by the designer — for example, overpressuring it by isolation of the relief valve. If designers do fit isolation valves below relief valves, then they are designing an error-prone plant. (Nevertheless, in certain cases, when the consequences of isolation are not serious, isolation of a relief valve may be acceptable.[7])

Another error made by vessel designers is inadvertently providing pockets into which water can settle. When the vessel heats up and hot oil comes into contact with the water, it may vaporize with explosive violence. In these cases the action required is better education of the designers, who were probably not aware of the hazard.

8.5 The Sellafield leak

A cause célèbre in 1984 was a leak of radioactive material into the sea from the British Nuclear Fuels Limited (BNFL) plant at Sellafield, Cumbria. It was the subject of two official reports[9,10] which agreed that the discharge was due to an operating error, though it is not entirely clear whether the error was due to a lack

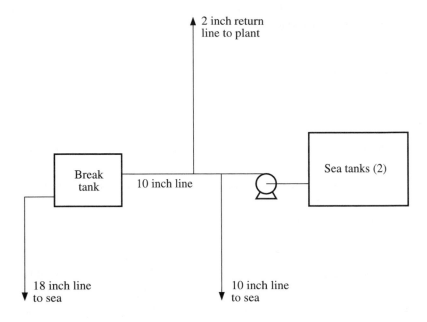

Figure 8.3 Simplified line diagram of the waste disposal system at Sellafield

of communication between shifts, poor training or wrong judgement. However, both official reports failed to point out that the leak was the result of a simple design error, that would have been detected by a hazard and operability study[1], if one had been carried out.

As a result of the operating error some material which was not suitable for discharge to sea was moved to the sea tanks (Figure 8.3). This should not have mattered as BNFL thought they had 'second chance' design, the ability to pump material back from the sea tanks to the plant. Unfortunately the return route used part of the discharge line to sea. The return line was 2 inches diameter, the sea line was 10 inches diameter, so solids settled out in the sea line where the linear flow rate was low and were later washed out to sea. The design looks as if it might have been the result of a modification. Whether it was or not, it is the sort of design error that would be picked up by a hazard and operability study.

The authors of the official reports seem to have made the common mistake of looking for culprits instead of looking for ways of changing the work situation, in this case by improving the design process.

Reference 1 describes many other accidents that could have been prevented by Hazop.

8.6 Other design errors

8.6.1 Breathing apparatus

A report[14] described several failures of the compressed air supply to breathing apparatus or compressed air suits. All the incidents could have been prevented by better design:

- A plunger in a quick-disconnect valve in a compressed air line was reversed. It should have been designed so that incorrect assembly was impossible.
- Another quick-disconnect valve came apart in use. A manufacturer's stamp on one of the components prevented it holding securely.
- A nozzle broke as the result of excessive stress.

8.6.2 Stress concentration

A non-return valve cracked and leaked at the 'sharp notch' shown in Figure 8.4(a) (page 162). The design was the result of a modification. The original flange had been replaced by one with the same inside diameter but a smaller outside diameter. The pipe stub on the non-return valve had therefore been turned down to match the pipe stub on the flange, leaving a sharp notch. A more knowledgeable designer would have tapered the gradient as shown in Figure 8.4(b) (page 162).

The detail may have been left to a craftsman. Some knowledge is considered part of the craft. We should not need to explain it to a qualified craftsman. He might resent being told to avoid sharp edges where stress will be concentrated. It is not easy to know where to draw the line. Each supervisor has to know the ability and experience of his team.

At one time church bells were tuned by chipping bits off the lip. The ragged edge led to stress concentration, cracking, a 'dead' tone and ultimately to failure[15].

8.6.3 Choice of materials

A dozen Cessna aircraft crashed in a period of 29 months, killing more than 29 people. In every case corrosion of the exhaust system was known or suspected to have occurred. After another crash in 1999 a Cessna spokesman was quoted in the press as saying that there was no fault in the design of the aircraft. 'It is a feature of the material which has shown that it does not take the wear over a number of years ... It was absolutely a viable design and they were viable materials for the aircraft.'[23] Doesn't the designer choose the materials of construction?

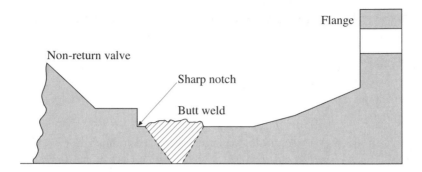

Figure 8.4(a) Turning reduced the thickness of the stub on the non-return (check) valve but left a sharp notch. Stress concentration led to failure.

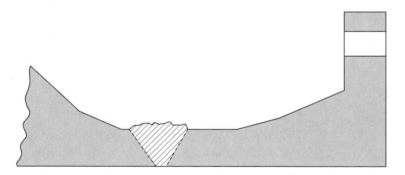

Figure 8.4(b) A safer solution

8.7 Conceptual shortcomings

Most of the incidents described so far have been mistakes: the designers were unaware of specific hazards, good practice or the practicalities of plant operation. Many of these designers would have benefited from a period as maintenance engineers or members of the start-up team on the plants they had designed. Ideally every design engineer should have worked on operating plant.

In addition, many designers are unaware of a whole range of design possibilities. Foremost amongst these is the concept of inherently safer design. Instead of controlling hazards by adding on protective equipment we can often avoid the hazard. Thus we can use so little hazardous material that it hardly matters if it all leaks out, we can use a safer material instead or we can use the hazardous material in the least hazardous form. We can often simplify the design so that there are fewer opportunities for errors and less equipment to fail (see Section 6.7.1, page 120). Reference 2 describes many examples of what has been or might be done and suggests procedures that can help us to identify possibilities. Very substantial reductions in inventory are sometimes possible.

The following example illustrates the concept. A potentially hazardous chemical reaction is the manufacture of ethylene oxide (EO) by oxidizing ethylene with pure oxygen, close to the explosive limit. However, the reaction takes place in the vapour phase and so the inventory in the reactor is small. When explosions have occurred, they have been localized incidents. The biggest hazard on many older EO plants is not the ethylene/oxygen mixture in the reactor but the method used for cooling it. The reactor tubes are surrounded by about 400 tonnes of boiling paraffin under pressure. Any major leak could produce a devastating vapour cloud explosion, similar to the one that occurred at Flixborough in 1974. Later plants use water under pressure instead of paraffin as the coolant[16].

Surveys have shown that most safety advisers are now familiar with the concept of inherently safe design. While most designers are aware of it they have a poor understanding of its scope and benefits. Most senior managers are unaware. Inherently safer designs require a major change in the design process, more time to consider alternatives in the early stages of design, and therefore they will not become widespread until senior managers become more aware and actively encourage their use.

Psychologists (and others) often say that we have reached the limit of what can be done by changes in design to make plants safer and we now need to concentrate on changing behaviour. This is not true. The best companies may be near the limit of what can be done by adding on protective equipment but the potentialities of inherently safer designs have still to be grasped. And, as the examples in this book show, many companies are far from using the best available designs.

There are many reasons why inherently safer designs have been adopted much more slowly than techniques such as Hazop and QRA[2,17]. One possible reason is that designers and others have a mind-set (see Section 4.4, page 88) that any technique of value must have a certain degree of complexity. Perhaps my catch-phrase, 'What you don't have, can't leak', sounds more like a gimmick than a recipe for action. Could it be a turn-off rather than a spark?

Other shortcomings of some designers are:

Failure to consider options for the future
When comparing alternative designs we should favour those that make further developments possible.

Excessive debottlenecking
It draws resources away from innovation and delays the date when brand new plants are built. New plants, using new technology, can make greater improvements in efficiency, energy saving, safety and protection of the environment[18].

Poor liaison with research chemists, especially on scale-up
According to Basu[19] the chemists do not know what information will be required and cannot take it into account when planning their experiments. He lists questions that should be asked.

Reluctance to learn from failures
According to Petrowski[20] (writing in 1994), 'By the late nineteenth and early twentieth century the increasing dramatic successes of engineers drove from the literature and, apparently, the mind of the engineering profession the failures and mistakes they then seemed doomed to repeat. Judging from the paucity of professional literature on failures (as opposed to failures themselves) in the first three-quarters of the present century, it was unfashionable at best, and unprofessional at worst, to deal too explicitly with the errors of engineering. This is unfortunate, but itself appears to be an error that is now being rectified.'

Reluctance to consider possible failure modes
According to Zetlin[21], 'Engineers should be slightly paranoiac during the design stage. They should consider and imagine that the impossible could happen. They should not be complacent and secure in the mere realization that if all the requirements in the design handbooks and manuals have been satisfied, the structure will be safe and sound.'

Poor liaison with other design sections
There was an example in Section 8.3.3 (page 159). Here is another. A plant extension was provided with its own control unit, located on a convenient site (electrical area classification Zone 2) a few metres from the main structure. It contained electrical equipment not suitable for use in a Zone 2 area and the electrical section of the design department therefore arranged for the equipment to placed in a metal box and purged with nitrogen. They did not ask if the control unit had to be in a Zone 2 area. That was not their job. Their job was to provide equipment suitable for the area classification already agreed. If they had queried the choice of site the control unit could have been moved to a safe area only a few metres away.

Not only was the blanketing expensive but its design was poor. It failed in its task and an explosion occurred[22].

Poor liaison with operators

If some designers fail to talk to their colleagues, more fail to discuss the design with those who will have to operate the plant. Many control system designers, for example, have never worked on a plant and may expect operators to look at various display pages at frequent intervals. Those who have operating experience will know that the same page may be left on display for long periods and will therefore provide read-only displays which, like old-fashioned chart recorders, will provide a continuous display of trends. Some companies appoint the commissioning manager early in design and he is based in the design department or contractor's office, to provide an operating input. One of the advantages of Hazop is that it forces different design sections and the commissioning manager to talk to each other, but unfortunately rather late in design.

Looking for alternatives in the wrong order

As already discussed, we should remove hazards when possible by inherently safer design. If we cannot do so, then we should add on protective equipment to keep the hazards under control. Only when that is not possible, should we depend on procedures. Unfortunately the default action of many designers, when they discover a hazard, is to act in the reverse order. They first propose a procedural solution. If they have been convinced that this is impractical, they propose adding on protective equipment. Only rarely do they look for ways of removing the hazard.

The fundamental shortcomings described in this section will not be overcome until senior managers recognize the need to do so and make the necessary organizational changes.

8.8 Problems of design contractors

Design contractors who are asked for a cheap design are in a quandary. Do they leave out safety features and safety studies such as Hazop, which are desirable but perhaps not essential, in order to get the contract? Obviously there is a limit beyond which no reputable contractor will go but with this reservation perhaps they should offer a minimum design and then say that they recommend that certain extra features and studies are added.

8.9 Domestic accidents

A report on domestic accidents[11] said, 'A more caring society would be the ideal but there is no way of obliging people to be more caring.' They therefore gave examples of ways in which better design has reduced or could reduce accidents in or near the home. For example:

- Since 1955 new paraffin heaters have had to be self-extinguishing when tipped over. Since then these heaters have caused fewer fires.
- Many burns are the result of falls onto electric fires. They could be prevented by using fires without exposed heating elements. Since 1964 new children's nightdresses have had to have increased resistance to flame spread. Since then fire deaths have fallen.
- Many drownings due to illness or drunkenness could be prevented by providing railings or closing paths at night.
- Many drownings of small children could be prevented by making the edges of ponds shallow, filling in disused reservoirs and canals or closing access through derelict land.

The report concludes that, 'Environmental and product design is the most reliable long-term means of accident prevention ... Already the design professions seem to take seriously designing for the physically handicapped who are otherwise active. Maybe they will soon see the need to design for the infirm, the mentally ill, drunks, drug addicts, neglected children, etc., or indeed anyone who is tired or under some form of stress.'

References in Chapter 8

1. Kletz, T.A., 1999, *Hazop and Hazan — Identifying and Assessing Process Industry Hazards*, 4th edition (Institution of Chemical Engineers, Rugby, UK).
2. Kletz, T.A., 1998, *Process Plants: A Handbook of Inherently Safer Design* (Taylor & Francis, Philadelphia, PA, USA).
3. *Boiler Explosions Act 1882 and 1890 — Preliminary Enquiry No. 3453*, 1969 (HMSO, London, UK).
4. Field, M., 1984, *Health and Safety at Work*, 6(12): 16.
5. Chapanis, A., 1965, *Man-machine Engineering*, page 42 (Tavistock Publications, London, UK).
6. Kletz, T.A., 2001, *Learning from Accidents*, 3rd edition, Chapter 16 (Butterworth-Heinemann, Oxford, UK).
7. Kletz, T.A., 1996, *Dispelling Chemical Industry Myths*, 3rd edition, items 2 and 5 (Taylor & Francis, Philadelphia, PA, USA).
8. *The Bulletin, The Journal of the Association for Petroleum Acts Administration*, April 1971.

9. *The Contamination of the Beach Incident at BNFL Sellafield*, 1984 (Health and Safety Executive, London, UK).

10. *An Incident Leading to Contamination of the Beaches near to the BNFL Windscale and Calder Works*, 1984 (Department of Environment, London, UK).

11. Poyner, B., 1980, *Personal Factors in Domestic Accidents* (Department of Trade, London, UK).

12. Foley, P., 1984, *Science Dimension*, 16(6): 13.

13. Forsberg, C.W., 1990, Passive and inherent safety technology for lightwater nuclear reactors, *AIChE Summer Meeting, San Diego, CA, USA, August 1990*.

14. *Operating Experience Weekly Summary*, 1998, No. 98–52, page 7 (Office of Nuclear and Facility Safety, US Department of Energy, Washington, DC, USA).

15. *Endeavour*, 1979, 3(1): 19.

16. Kletz, T.A., 1998, *Process Plants: A Handbook for Inherently Safer Design*, Section 4.1.2 (Taylor & Francis, Philadelphia, PA, USA).

17. Kletz, T.A., 1999, *Process Safety Progress*, 18(1): 64.

18. Malpas, R., 1986, in Atkinson, B., *Research and Innovation for the 1990s*, page 28 (Institution of Chemical Engineers, Rugby, UK).

19. Basu, P.K., 1988, *Chemical Engineering Progress*, 94(9): 75.

20. Petrowski, H., 1994, *Design Paradigms*, page 98 (Cambridge University Press, Cambridge, UK).

21. Zetlin, L., 1988, quoted by Petrowski, H., 1994, *Design Paradigms*, page 3 (Cambridge University Press, Cambridge, UK).

22. Kletz, T.A., 2001, *Learning from Accidents*, 3rd edition, Chapter 2 (Butterworth-Heinemann, Oxford, UK).

23. Uhlig, R., *Daily Telegraph*, 4 September 1999.

Some accidents that could be prevented by better construction

'... on the one hand the most brilliant workmanship was disclosed, while on the other hand it was intermingled with some astonishing carelessness and clumsiness.'
Flinders Petrie (describing the pyramids)[3]

'How very little, since things were made,
Things have altered in the building trade.'
Rudyard Kipling, *A Truthful Song*

This chapter describes some accidents which were the result of construction errors; in particular, they resulted from the failure of construction teams to follow the design in detail or to do well, in accordance with good engineering practice, what had been left to their discretion. The reasons for the failures may have been lack of training or instructions or a lack of motivation, but the actions needed are the same: better inspection during and after construction in order to see that the design has been followed and that details left to the discretion of the construction team have been carried out in accordance with good engineering practice, and perhaps better training.

The construction error that led to the collapse of the Yarra bridge has already been discussed (see Section 3.3.5, page 57).

9.1 Pipe failures

The following examples of construction errors which led to pipe failures (or near-failures) are taken from a larger review of the subject[1]. Many of the failures occurred months or years after the construction errors.

- A temporary support, erected to aid construction, was left in position.
- A plug, inserted to aid pressure testing, was left in position. It was not shown on any drawing and blew out 20 years later.
- Pipes were inadequately supported, vibrated and failed by fatigue. (The operating team could also have prevented the failure, by seeing that the pipe was adequately supported.)

- A construction worker cut a hole in a pipe at the wrong place and, discovering his error, patched the pipe and said nothing. The patch was then covered with insulation. As the weld was not known to exist it was not radiographed. It was substandard and corroded. There was a leak of phosgene and a man was nearly killed.
- A new line had the wrong slope so the contractors cut and welded some of the hangers. They failed. Other hangers failed due to incorrect assembly and absence of lubrication.
- Two pipe ends, which were to be welded together, did not fit exactly and were welded with a step between them.
- On several occasions bellows have failed because the pipes between which they were installed were not lined up accurately. The men who installed them apparently thought that bellows can be used to take up misalignment in pipes. In fact, when bellows are used, pipes should be aligned more, not less, accurately than usual (see Section 7.10, page 149).
- Pipes have been fixed to supports so firmly that they were not free to expand, and tore in trying to do so.
- When a pipe expanded, a branch came in contact with a girder, on which the pipe was resting, and was knocked off. The pipe was free to expand 125 mm (5 inches) but actually expanded 150 mm (6 inches). Did the construction team know that the pipe would be used hot and would therefore expand and, if so, by how much?
- Insufficient forged T-pieces were available for a hot boiler feed water system, so three were improvised by welding together bits of pipe. Four years later they failed.
- Dead-ends have been left in pipelines in which water and/or corrosion products have accumulated. The water has frozen, splitting the pipe, or the pipe has corroded.
- On many occasions construction teams have used the wrong material of construction. Often the wrong material has been supplied, but the construction team did not carry out adequate checks. The responsibility lies with the construction management rather than the workers.

As an example of the last bullet, an ammonia plant was found to have the following faults, none of which were detected during construction and all of which caused plant shutdowns[2]:

- Turbine blades were made from the wrong grade of steel.
- The bolts in the coupling between a turbine and a compressor had the wrong dimensions.

- There was a machining error in another coupling.
- Mineral-filled thermocouples were filled with the wrong filler.
- The rivets in air compressor silencers were made from the wrong material.

Other faults detected in time included:

- Defects in furnace tubes.
- Blistering caused by welding fins onto furnace tubes.
- Some furnace tubes made from two different grades of steel.
- Some furnace tube bends made from the wrong grade of steel.

Following similar experiences, in the 1970s and 1980s many companies introduced positive materials identification programmes. If use of the wrong grade of material could have adverse results, such as premature failure or corrosion, all construction and maintenance material, from welding rods to complete items, was analysed to check that it was made from the material specified. Many of these programmes were abandoned when suppliers achieved quality certification. Since a minor error — from the supplier's point of view — can have disastrous results, is this wise?

A list of points to check during and after construction is included in Reference 1. Many of them are not included in any code of practice as they are considered obvious; they are good engineering practice. Probably no code, for example, says that pipes must not be fixed so firmly that they cannot expand. Nevertheless, this has been done. When inspecting a plant after construction we have to look out for errors that no-one has ever specifically forbidden.

9.2 Miscellaneous incidents

9.2.1 Contractors made it stronger

A compressor house was designed so that the walls would blow off if an explosion occurred inside. The walls were to be made from lightweight panels secured by pop rivets.

The construction team decided that this was a poor method of securing the wall panels and used screws instead. When an explosion occurred in the building the pressure rose to a much higher level than intended before the walls blew off, and damage was greater that it need have been. In this case a construction engineer, not just a construction worker, failed to follow the design. He did not understand, and had probably not been told, the reason for the unusual design. Section 3.3.4 (page 56) describes another similar incident.

9.2.2 Contractors exceeded authorization

An explosion occurred in a new storage tank (Figure 9.1), still under construction. The roof was blown off and landed, by great good fortune, on one of the few pieces of empty ground in the area. No-one was hurt. Without permission from the operating team, and without their knowledge, the construction team had connected up a nitrogen line to the tank. They would not, they said, have connected up a product line but they thought it would be quite safe to connect up the nitrogen line. Although the contractors closed the valve in the nitrogen line (Figure 9.1), it was leaking and a mixture of nitrogen and flammable vapour entered the new storage tank. The vapour mixed with the air in the tank and was ignited by a welder who was completing the inlet piping to the tank.

The contractors had failed to understand that:

- The vapour space of the new tank was designed to be in balance with that of the existing tank, so the nitrogen will always be contaminated with vapour.
- Nitrogen is a process material, it can cause asphyxiation, and it should be treated with as much care and respect as any other process material.
- No connection should be made to existing equipment without a permit-to-work, additional to any permit issued to construct new equipment. Once new equipment is connected to existing plant it becomes part of it and should be subject to the full permit-to-work procedure.

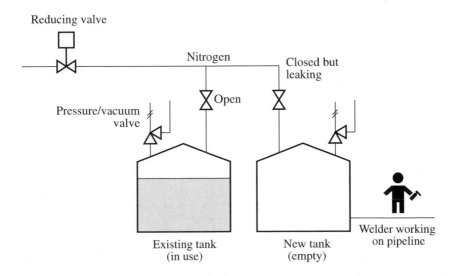

Figure 9.1 Arrangement of nitrogen lines on two tanks. Connection of the new tank to the nitrogen system led to an explosion.

9.2.3 Six errors in twelve inches

Six errors, any one of which could have led to a major incident, were found in a 12-inch length of equipment following a minor modification: insertion of a flanged valve made from a socket weld valve, two socket weld flanges and two short pipes.

- The flanges were made from the wrong grade of steel, a grade unsuitable for welding.
- They were machined from bar, though forgings were specified.
- The bar had cracked during heat treatment and this was not discovered until the flanges failed in service.
- The valve was also made from the wrong grade of steel.
- The post-welding heat treatment was omitted.
- Pressure testing of the completed assembly was ineffective or not carried out.

In engineering we communicate by quoting specifications. This is ineffective if the people we are talking to do not understand them or do not realize the importance of following them.

9.2.4 A few small holes left out

A new atmospheric pressure tank was constructed for the storage of water. The level was measured by a float suspended inside a vertical tube the full height of the tank. Holes were drilled in the tube to make sure that the level inside it was always the same as the level outside it.

One day the tank was overfilled and the roof came off although the level indicator said it was only about half full. It was then found that the construction team had not drilled any holes in the top half of the tube. As the level in the tank rose the air in the tube was compressed and the level in the tube rose less than the level in the tank.

Perhaps the tank was inspected less thoroughly than usual, as it was 'only a water tank'.

9.2.5 Poor co-ordination

Some contractors, welding near the air inlet to a ventilation system, accidentally set fire to some insulation. Fumes were sucked into the building. The fire was soon extinguished; no-one was harmed by the fumes but a $17,000 experiment was ruined. No-one considered the effects the contractors might have on the ventilation[4].

9.3 Prevention of construction errors

In the opening paragraph of this chapter I suggested that errors by construction teams are best detected by detailed inspection during and after construction. Who should carry out the inspection? The checks made by construction inspectors in the past are clearly not sufficient.

The team who will start up and operate the plant have an incentive to inspect the plant thoroughly, as they will suffer the results of any faults not found in time. The designers, though they may not spot their own errors, may see more readily than anyone else when their intentions have not been followed. The inspection of the plant during and after construction should therefore be carried out by the start-up team assisted by one or more members of the design team.

Could we reduce the number of construction errors by taking more trouble to explain to construction workers the nature of the materials to be handled and the consequences of not following the design and good practice? Usually little or no attempt is made to carry out such training and many construction engineers are sceptical of its value, because of the itinerant nature of the workforce. Nevertheless, perhaps it might be tried. Certainly, there is no excuse for not telling construction engineers and supervisors why particular designs have been chosen, and thus avoiding errors such as some of those described in Sections 9.1 and 9.2.

Obviously, because of the nature of the task, it is difficult to prevent construction errors by changes in design and the approach must be mainly a software one — better inspection and perhaps training. However we should, when possible, avoid designs which are intolerant of construction errors. Bellows, for example, may fail if the pipes between which they are placed are not lined up accurately. Fixed piping can be forced into position, but not bellows. Bellows are therefore best avoided, when hazardous materials are handled, and flexibility obtained by incorporating expansion loops in the piping.

Similarly, errors similar to that shown in Figure 8.2 (page 158) can be avoided by designing the equipment so that it is impossible to assemble it incorrectly.

This chapter is not intended as an indictment of construction teams. They have a different background to those who design and operate plants. What is obvious to designers and operators is not obvious to them.

References in Chapter 9

1. Kletz, T.A., 2001, *Learning from Accidents*, 3rd edition, Chapter 16 (Butterworth-Heinemann, Oxford, UK).
2. Kolff, S.W. and Mertens, P.R., 1984, *Plant/Operations Progress*, 3(2): 117.
3. Petrie, F., 1893, *Ten Years Digging in Egypt* (Religious Tract Society, London, UK).
4. *Co-ordinating Construction/Maintenance Plans with Facility Manager may Deter Unexpected Problems and Accidents, Safety Note DOE/EH–0127*, 1990 (US Department of Energy, Washington, DC, USA).

Some accidents that could be prevented by better maintenance

10

'*The biggest cause of breakdowns is maintenance.*'
Anon

This chapter describes some accidents which were the result of maintenance errors. Sometimes the maintenance workers were inadequately trained, sometimes they took short cuts, sometimes they made a slip or had a moment's lapse of attention. Sometimes it is difficult to distinguish between these causes, or more than one was at work. As with construction errors, it is often difficult or impossible to avoid them by a change in design and we are mainly dependent on training and inspection. However, designs which can be assembled incorrectly should be avoided, as should equipment such as bellows (see Section 7.10, page 149, and Section 9.3, page 173), which are intolerant of poor quality maintenance.

10.1 Incidents which occurred because people did not understand how equipment worked

(a) People have been injured when dismantling diaphragm valves because they did not realize that the valves can contain trapped liquid (Figure 10.1, page 176). Here again, a hardware solution is often possible — liquid will not be trapped if the valves are installed in a vertical section of line.

(b) On a number of occasions men have been asked to change a temperature measuring device and have removed the whole thermowell. One such incident, on a fuel oil line, caused a serious refinery fire[1].

(c) On several occasions, when asked to remove the actuator from a motorized valve, men have undone the wrong bolts and dismantled the valve[2,3]. One fire which started in this way killed six men. On other occasions trapped mechanical energy, such as a spring under pressure, has been released.

A hardware solution is possible in these cases. Bolts which can safely be undone when the plant is up to pressure could be painted green; others could be painted red[4]. A similar suggestion is to use bolts with recessed heads and fill the heads with lead if the bolts should not be undone when the plant is up to pressure.

175

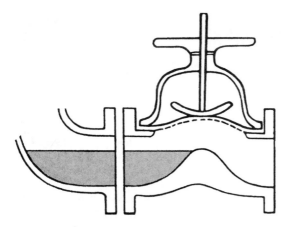

Figure 10.1 Liquid trapped in a diaphragm valve. This can be avoided by locating the valve in a vertical line.

(d) As discussed in Section 5.3 (page 109), a fitter was asked to dismantle a valve. The design was unfamiliar and as a result there was a sudden release of trapped pressure. A more safety-conscious fitter might have got a spare valve out of the store and dismantled it first to make sure he knew how the valve was constructed. However, this would have caused delay and would have meant admitting that he did not know the construction of the valve. The culture of the workplace caused him to press on. The company had made a big investment in behavioural safety training. People worked more safely and the accident rate, both lost-time and minor, had fallen dramatically, but the behavioural training had not recognized and tackled the macho culture that led to the accident.

(e) Figure 10.2 illustrates another incident. Cables were suspended from the roof of a large duct by wire hangers with a hook at each end. Both ends of the hangers were hooked over a steel girder. The cables had to be moved temporarily so that some other work could be carried out. The three men who moved the cables decided to hook the hangers over the top of a bracket that was fixed to the adjoining wall. There would have been no problem if they had hooked both ends of each hanger onto the bracket. However, they hooked only one end of each hanger onto the bracket and hooked the other end onto the hanger itself, as shown in the lower part of Figure 10.2. They did not realize that this doubled the weight on the end fixed to the wall. The weight of the cables straitened the end of one of the hangers. The adjacent hangers then failed. Altogether 60 m (200 feet) of cable fell 5 m (15 feet). One man was injured, fortunately not seriously[13].

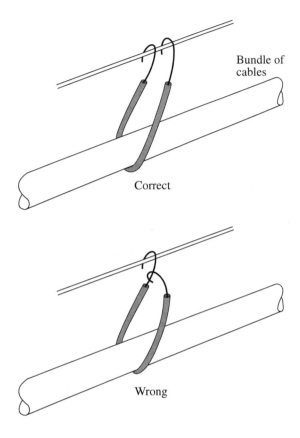

Bundle of
cables

Correct

Wrong

Figure 10.2 Two ways of supporting a bundle of cables with a wire hanger

How far do we need to go in giving detailed instructions about every aspect of every job, especially to trained craftsmen? In this case, the hazard is one that many people might overlook. Each supervisor must know his team and their abilities. There is a move today towards self-managed groups. Does this make incidents such as the one just described more likely?

A hardware solution is possible: the hangers could be made strong enough to support the weight even if they are used as shown in the lower part of Figure 10.2.

(f) A cylinder containing gas at 2200 psi (150 bar) was connected to a filter and then to a reducing valve which lowered the pressure to 300 psi (20 bar). The drawing incorrectly showed the filter rated for 300 psi. A craftsman therefore installed a filter suitable for this pressure. It ruptured as the isolation valve on the cylinder was opened. Many craftsmen would have spotted the unsuitability of the filter as soon as they started work[14].

(g) A ventilation system failed because a craftsman put too much grease on the air-flow sensors. The site originally employed its own craftsman but as the result of budget cuts it changed to an outside organization. The site could no longer rely on the 'skill of the craft' now that repairs were carried out by a succession of strangers. A number of similar incidents were reported[15].

10.2 Incidents which occurred because of poor maintenance practice

(a) It is widely recognized that the correct way to break a bolted joint is to slacken the bolts and then wedge the joint open on the side furthest from the person doing the job. If there is any pressure in the equipment, then the leakage is controlled and can either be allowed to blow off or the joint can be remade. Nevertheless accidents occur because experienced men undo all the bolts and pull a joint apart.

For example, two men were badly scalded when they were removing the cover from a large valve on a hot water line, although the gauge pressure was only 9 inches of water (0.33 psi or 0.023 bar). They removed all the nuts, attached the cover to a hoist and lifted it.

(b) On many occasions detailed inspections of flameproof electrical equipment have shown that many fittings were faulty — for example, screws were missing or loose, gaps were too large, glasses were broken.

This example illustrates the many-pronged approach necessary to prevent many human error accidents:

Hardware
Flameproof equipment requires careful and frequent maintenance. It is sometimes, particularly on older plants, used when Zone 2 equipment — cheaper and easier to maintain — would be adequate. Flameproof equipment requires special screws and screwdrivers but spares are not always available.

Training
Many electricians do not understand why flameproof equipment is used and what can happen if it is badly maintained.

Inspection
Experience shows that it is necessary to carry out regular inspections or audits of flameproof equipment if standards are to be maintained. Often equipment at ground level is satisfactory but a ladder discloses a different state of affairs!

(c) When a plant came back on line after a turnaround it was found that on many joints the stud bolts were protruding too far on one side and not enough on the other, so that some nuts were secured by only a few threads.

On eight-bolt joints the bolts were changed one at a time. Four-bolt joints were secured with clamps until the next shutdown.

(d) A young engineer was inspecting the inside of a 1.5 m (60 inches) diameter gas main, wearing breathing apparatus supplied from the compressed air mains. He had moved 60 m (200 feet) from the end when his face mask started to fill with water. He pulled it off, held his breath, and walked quickly to the end.

He discovered that the air line had been connected to the bottom of the compressed air main instead of the top. As a young engineer, with little experience, he had assumed that the 'safety people' and the factory procedures would do all that was required and that he could rely on them.

It is, of course, the responsibility of those who issue the entry permit and those who look after the breathing apparatus to make sure everything is correct, but men should nevertheless be encouraged, before entering a vessel, to carry out their own checks.

A hardware solution is possible in this case. If breathing apparatus is supplied from cylinders or from portable compressors instead of from the factory compressed air supply, contamination with water can be prevented.

Of course, no system is perfect. Those responsible may fail to supply spare cylinders, or may not change them over in time, or may even supply cylinders of the wrong gas. Compressors may fail or be switched off.

We are dependent on procedures to some extent, whatever the hardware. We depend on training to make sure people understand the procedures and on inspections to make sure people follow the procedures. But some designs provide more opportunities for error than others and are therefore best avoided. On the whole piped air is more subject to error than air from cylinders or a compressor.

(e) The temperature controller on a reactor failed and a batch overheated. It was then found that there was a loose terminal in the controller. The terminal was secured by a bolt which screwed into a metal block. The bolt had been replaced by a longer one which bottomed before the terminal was tight.

(f) Companies have procedures for examining all lifting gear at regular intervals. Nevertheless, corroded or damaged wire ropes and slings often fail in service[16]. Being small, they are easily mislaid, reported missing and then found and used again. A good practice is to fix a coloured tag to each item after examination and to change the colour every six months. Everyone can then see if an item is overdue for examination.

(g) On many occasions someone has broken the wrong joint, changed the wrong valve or cut open the wrong pipeline. Sometimes the job has been shown to the fitter who has gone for his tools and then returned to the wrong joint or valve; sometimes a chalk mark has been washed off by rain or the fitter has been misled by a chalk mark left from an earlier job; sometimes the job has been described but the description has been misunderstood. Equipment which is to be maintained should be identified by a numbered tag (unless it is permanently labelled) and the tag number shown on the permit-to-work[17]. Instructions such as 'The pump you repaired last week is giving trouble again' lead to accidents.

Sometimes the process team members have identified the equipment wrongly. To stop a steam leak a section of the steam line was bypassed by a hot tap and stopple. After the job was complete it was found that the steam leak continued. It was then found that the leak was actually from the condensate return line to the boiler which ran alongside the steam line. The condensate was hot and flashed into steam as it leaked. Several people then said that they had thought that the steam seemed rather wet[11]! This incident could be classified as a 'mind-set' (see Section 4.4, page 88). Having decided that the leak was coming from the steam main, everyone closed their minds to the possibility that it might be coming from somewhere else.

For another incident see Section 11.1, first item, page 187.

10.3 Incidents due to gross ignorance or incompetence

As in Section 3.4 on page 62, some incidents have been due to almost unbelievable ignorance of the hazards.

A shaft had to be fitted into a bearing in a confined space. It was a tight fit so the men on the job decided to cool the shaft and heat the bearings. The shaft was cooled by pouring liquefied petroleum gas onto it while the bearing was heated with an acetylene torch[5]!

A fitter was required to remake a flanged joint. The original gasket had been removed and the fitter had to obtain a new one. He selected the wrong type and found it was too big to fit between the bolts. He therefore ground depressions in the outer metal ring of the spiral wound gasket so that it would fit between the bolts (Figure 10.3)! As if this were not bad enough, he ground only three depressions, so the gasket did not fit centrally between the bolts but was displaced 12 mm to one side.

The fitter's workmanship is inexcusable but the possibility of error in selecting the gasket was high as four different types of joint were used on the plant concerned.

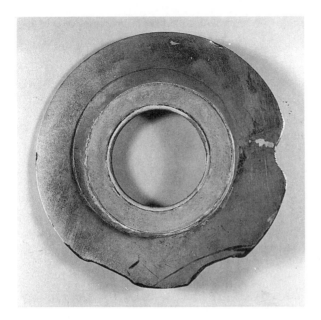

Figure 10.3 A spiral wound gasket was ground away to make it fit between the bolts of a joint

10.4 Incidents which occurred because people took short cuts

Many accidents have occurred because permit-to-work procedures were not followed. Often it is operating teams who are at fault (see Chapter 10) but sometimes the maintenance team are responsible.

Common faults are:

- carrying out a simple job without a permit-to-work;
- carrying out work beyond that authorized on the permit-to-work;
- not wearing the correct protective clothing.

To prevent these incidents we should:

- train people in the reasons for the permit system and the sort of accidents that occur if it is not followed. There is, unfortunately, no shortage of examples and some incidents are described and illustrated in the Institution of Chemical Engineers' safety training package on *Safer Maintenance*[6];
- check from time to time that the procedures are being followed (see Chapter 5).

The following are examples of accidents caused by short-cutting:

(a) A fitter and an apprentice were affected by gas while replacing a relief valve in a refinery. They had not obtained a permit-to-work. If they had, it would have stipulated that breathing apparatus should be worn[7]. The fitter's 'excuse' was that he was unaware that the equipment was in use. The fitter and apprentice both smelled gas after removing blanks but took no notice. Many accidents could be avoided if people responded to unusual observations.

(b) Several accidents have occurred because fitters asked for and received a permit to examine the bearings (or some other external part) on a pump or other machine and later decided that they needed to open up the machine. They then did so without getting another permit and a leak occurred. If a permit is issued for work on bearings, the process lines may not have been fully isolated and the machine may not have been drained.

(c)   A permit was issued for a valve to be changed on a line containing corrosive chemicals. It stated that gloves and goggles must be worn. The fitter did not wear them and was splashed with the chemical.

The line had been emptied and isolated by locked valves but some of the corrosive liquid remained in the line.

At first sight this accident seems to be due to the fault of the injured man and there is little that management can do to prevent it, except to see that rules are enforced. However, examination of the permit book showed that every permit was marked 'Gloves and goggles to be worn', though many of them were for jobs carrying no risk. The maintenance workers therefore ignored the instruction and continued to ignore it even on the odd occasion when it was really necessary.

If the fitter had been better at putting his case he would have said, 'Why didn't you tell me that for once goggles were necessary? Why didn't you write on the permit, "Gloves and goggles to be worn and this time I mean it"?'

Do not ask for more protective clothing than is necessary.

Ask only for what is necessary and then insist that it is worn.

Why did the operating team ask for more protective clothing than was necessary? I suspect that at some time in the past a supervisor had asked for too little, someone had been injured and the supervisor had been blamed. All supervisors then started to ask for gloves and goggles every time. If we allow those who are responsible to us to use their discretion in borderline cases, then inevitably they will sometimes come to different decisions than those we might have come to ourselves. We should check their decisions from time to time and discuss with them the reasons for any we consider doubtful. But coming to a different decision does not justify blame.

(d) Many accidents have occurred because the operating team failed to isolate correctly equipment which was to be repaired. It is sometimes suggested that the maintenance workers should check the isolations and in some companies this is required.

In most companies however the responsibility lies clearly with the operating team. Any check that maintenance workers carry out is a bonus. Nevertheless they should be encouraged to check. It is their lives that are at risk.

On one occasion a large hot oil pump was opened up and found to be full of oil. The ensuing fire killed three men and destroyed the plant. The suction valve on the pump had been left open and the drain valve closed.

The suction valve was chain-operated and afterwards the fitter recalled that earlier in the day, while working on the pump bearings, the chain had got in his way. He picked it up and, without thinking, hooked it over the projecting spindle of the open suction valve!

This incident also involved a change of intention. Originally only the bearings were to be worked on. Later the maintenance team decided that they would have to open up the pump. They told the process supervisor but he said that a new permit was not necessary.

(e) A fitter was asked to fit a slip-plate between a pair of flanges. Finding it difficult to wedge them apart he decided, without consulting his supervisor or the process team, to fit the slip-plate between another pair of flanges. As they were on the other side of the isolation valve which isolated the joint he was supposed to break, a corrosive chemical sprayed onto his face. He was wearing goggles but knocked them off as he flung up his arms to protect his face.

This incident raises several questions. Did the fitter understand the nature of the liquid in the pipeline and the need to follow instructions precisely? Had he fitted slip-plates in the wrong position before? Had anyone had difficulty slip-plating this joint before? If so, had modifications, or slip-plating elsewhere, been considered? When the need for slip-plating can be foreseen during design, do designers ensure that access and flexibility are adequate?

10.5 Incidents which could be prevented by more frequent or better maintenance

All the incidents just described could be prevented by better management — that is, by better training, supervision, etc. Other incidents have occurred because of a misjudgement of the level or quality of maintenance required. Such incidents are rare in the oil and chemical industries, but occasionally one hears of a case in which routine maintenance or testing has been repeatedly postponed because of pressure of work.

For example, after a storage tank had been sucked in it was found that the flame arresters in the three vents were choked with dirt. Although scheduled for three-monthly cleaning they had not been cleaned for two years.

It may not matter if the routine cleaning of flame arresters is postponed for a week or a month or even if a whole cycle of cleaning is omitted, but if we continue to put off cleaning an accident in the end is inevitable.

In many cases, if we neglect safety measures, we take a chance and there may be no accident; in the case just described an accident was almost certain.

However, a change in design might have prevented the accident. The flame traps had to be unbolted and cleaned by maintenance workers. If they could have been secured without bolts, so that they could have been cleaned by process operators, it is less likely that cleaning would be neglected.

This simple incident illustrates the theme of this book: it is easy to talk of irresponsibility and lack of commitment and to urge engineers to conform to schedules. It is harder, but more effective, to ask why schedules are not followed and to change designs or methods so that they are easier to follow.

A book on railway boiler explosions[8] shows that in the period 1890–1920 the main reasons for these were poor quality maintenance or a deliberate decision to keep defective boilers on the road. Many of the famous railway engineers such as William Stroudley and Edward Fletcher come out badly. Great designers they may have been but — as mentioned in Section 5.3 (page 107) — their performance as maintenance engineers was abysmal; too much was left to the foremen.

Few failures were due to drivers tampering with relief valves or letting the water level in the boiler get too low. Locomotive superintendents, however, encouraged the view that many explosions were due to drivers tampering with relief valves. It is easy to blame the other man.

In 1966 a colliery tip collapsed at Aberfan in South Wales, killing 144 people, most of them children. The official report[9,10] showed that similar tips had collapsed before, though without serious consequences, and that the action needed to prevent collapse had been well known for many years.

Following a collapse in 1965 all engineers were asked to make a detailed examination of tips under their control. The Aberfan tip was inspected in the most perfunctory manner. The Inquiry criticized the engineer responsible 'for failing to exercise anything like proper care in the manner in which he purported to discharge the duty of inspection laid upon him'. Responsibility for tip inspection had been given to mechanical engineers instead of civil engineers.

10.6 Can we avoid the need for so much maintenance?

Since maintenance results in so many accidents — not just accidents due to human error but others as well — can we change the work situation by avoiding the need for so much maintenance?

Technically it is certainly feasible. In the nuclear industry, where maintenance is difficult or impossible, equipment is designed to operate without attention for long periods or even throughout its life. In the oil and chemical industries it is usually considered that the high reliability necessary is too expensive.

Often, however, the sums are never done. When new plants are being designed, often the aim is to minimize capital cost and it may be no-one's job to look at the total cash flow. Capital and revenue may be treated as if they were different commodities which cannot be combined. While there is no case for nuclear standards of reliability in the process industries, there may sometimes be a case for a modest increase in reliability.

Some railway rolling stock is now being ordered on 'design, build and maintain' contracts. This forces the contractor to consider the balance between initial and maintenance costs.

For other accounts of accidents involving maintenance, see Reference 12.

Afterthought

'Once correctly tightened, a simple line is painted on both nut and bolt, so it is easy to see if the bolt has loosened — the lines on nut and bolt fall out of alignment with loosening.'

'I saw plenty of high-tech equipment on my visit to Japan, but I do not believe that of itself this is the key to Japanese railway operation — similar high-tech equipment can be seen in the UK. Pride in the job, attention to detail, equipment redundancy, constant monitoring — these are the things that make the difference in Japan, and they are not rocket science ...'

'There have been many study tours by British railway officials to Japan with little follow up ... We should start applying the lessons or we should stop going on the study tours.'
James Abbott[18]

References in Chapter 10

1. *Petroleum Review*, October 1981, page 21.
2. Kletz, T.A., 1998, *What Went Wrong? — Case Histories of Process Plant Disasters*, 4th edition, Chapter 1 (Gulf Publishing Co, Houston, Texas, USA).
3. Kletz, T.A., in Fawcett, H.H. and Wood, W.S. (editors), 1982, *Safety and Accident Prevention in Chemical Operations*, Chapter 36 (Wiley, New York, USA).
4. Kletz, T.A., 1984, *The Chemical Engineer*, No. 406, page 29.
5. Cloe, W.W., April 1982, *Selected Occupational Fatalities Related to Fire and/or Explosion in Confined Workspaces as Found in Reports of OSHA Fatality/Catastrophe Investigation, Report No. OSHA/RP–82/002* (US Department of Labor).
6. Safety training packages 028 *Safer Work Permits* and 033 *Safer Maintenance* (Institution of Chemical Engineers, Rugby, UK).
7. *Petroleum Review*, April 1984, page 37.
8. Hewison, C.H., 1983, *Locomotive Boiler Explosions* (David and Charles, Newton Abbot, UK).
9. *Report of the Tribunal Appointed to Inquire into the Disaster at Aberfan on October 21st, 1966*, 1967, paragraph 171 (HMSO, London, UK).
10. Kletz, T.A., 2001, *Learning from Accidents*, 3rd edition, Chapter 13 (Butterworth-Heinemann, Oxford, UK).
11. Ramsey, CW., 1989, Plant modifications: maintain your mechanical integrity, *AIChE Loss Prevention Symposium, Houston, Texas, USA, April 1989*.
12. Health and Safety Executive, 1987, *Dangerous Maintenance: A Study of Maintenance Accidents in the Chemical Industry and How to Prevent Them* (HMSO, London, UK).
13. *Operating Experience Weekly Summary*, 1999, No. 99–08, page 1 (Office of Nuclear and Facility Safety, US Department of Energy, Washington, DC, USA).
14. *Operating Experience Weekly Summary*, 1998, No. 98–23, page 3 (Office of Nuclear and Facility Safety, US Department of Energy, Washington, DC, USA).
15. *Operating Experience Weekly Summary*, 1998, No. 98–06, page 11 (Office of Nuclear and Facility Safety, US Department of Energy, Washington, DC, USA).
16. *Operating Experience Weekly Summary*, 1998, No. 98–22, page 2 (Office of Nuclear and Facility Safety, US Department of Energy, Washington, DC, USA).
17. Kletz, T.A. in Grossel, S.S. and Crowl, D.A. (editors), 1995, *Handbook of Highly Toxic Materials and Management*, Chapter 11 (Dekker, New York, USA).
18. Abbott, J., 2001, *Modern Railways*, January, page i.

Some accidents that could be prevented by better methods of operation

'When safe behaviour causes trouble, a change occurs in the unsafe direction.'
K. Hakkinen[8]

This chapter describes some incidents which occurred as the result of errors by operating staff. It is not always clear which of the types of error discussed in Chapters 2–5 were involved. Often more than one was at work.

11.1 Permits-to-work

Accidents which occurred as the result of a failure to operate a permit-to-work system correctly are discussed briefly in Section 5.2.1 (page 103), while Section 10.4 (page 181) describes incidents which occurred because maintenance workers cut corners.

Here is another incident:

What happened	What should have happened
A permit was issued to connect a nitrogen hose to flange A (see Figure 11.1 on page 188), so that equipment could be leak tested, and then disconnect it when the test was complete.	Two permits should have been issued: one to connect the hose and a second one — when the time came — to disconnect it.
Flange A was not tagged.	It should have been tagged. As the job is done several times per year there should have been a permanent tag.
When the leak test was complete the process supervisor (who was a deputy) asked the lead fitter to disconnect the hose. He did not show him the job or check that it was tagged.	He should have shown him the job and checked that it was tagged.

The fitter who was asked to do the job — not the one who had fitted the hose — misunderstood his instructions and broke joint B. He was new to the works.	He should have been better trained.
The lead fitter signed off the permit without inspecting the job.	He should have inspected it.
The process supervisor accepted the permit back and started up the plant without inspecting the job.	He should have inspected it.
Toxic gas came out of the open joint B. Fortunately, no-one was injured.	

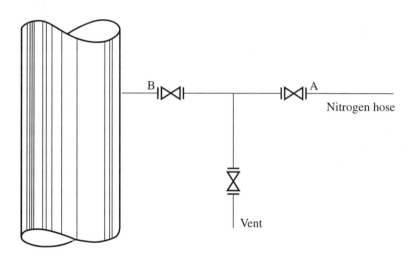

Figure 11.1 Joint B was broken instead of Joint A

There were at least eight 'human errors':
(1) and (2) Failure of two process supervisors to tag the job.
(3) Failure of the process supervisor to show the job to the second fitter.
(4) Failure of the lead fitter to inspect the completed job.
(5) Failure of the process supervisor to inspect the completed job.
(6) Failure of the manager to see that the deputy process supervisor was better trained.

(7) Failure of the maintenance engineer and foreman to see that the new fitter was better trained.

(8) Failure of the manager and maintenance engineer to monitor the operation of the permit system.

If any one of these errors had not been made, the accident would probably not have occurred. In addition the process supervisor issued one permit instead of two, though issuing two permits would probably not have prevented the accident.

In a sense, (8) includes all the others. If the manager and engineer had kept their eyes open before the incident and checked permits-to-work regularly, they could have prevented their subordinates from taking short cuts and recognized the need for further training.

Prevention of similar accidents depends on better training and monitoring. However, a hardware approach is possible.

If the nitrogen hose is connected by means of a clip-on coupling, the maintenance organization need not be involved at all and the nitrogen supply cannot be disconnected at the wrong point. Alternatively, since the nitrogen is used several times per year, it could be permanently connected by a double block and bleed valve system. The method actually used was expensive as well as providing opportunities for error.

Adoption of the hardware solution does not lessen the need for training and monitoring. Even if we prevent, by a change in design, this particular accident occurring again, failure to follow the permit system correctly will result in other accidents.

11.2 Tanker incidents

11.2.1 Overfilling

Road and rail tankers are often overfilled. If the tankers are filled slowly, the filler is tempted to leave the job for a few minutes. He is away longer than he expects to be and returns to find the tanker overflowing.

If the tankers are filled quickly, by hand, then the operator has only to be distracted for a moment for the tanker to be overfilled. Also, accurate metering is difficult. For these reasons most companies use automatic meters. The quantity to be put into the tanker is set on a meter which closes the filling valve automatically when this quantity has been passed. Unfortunately these meters do not prevent overfilling as:

- the wrong quantity may be set on the meter, either as the result of a slip or because of a misunderstanding of the quantity required;
- there may be a residue in the tanker left over from the previous load which the operator fails to notice;
- there may be a fault in the meter.

For these reasons many companies have installed high level trips in their tankers. They close the filling valve automatically when the tanker is nearly full. Although a separate level measuring device is required in every compartment of every tanker, they are relatively cheap.

11.2.2 Overfilling pressure tankers

When liquefied gases are transported, pressure tankers are used. Overfilling does not result in a spillage as the vapour outlet on the top of the tanker is normally connected to a stack or a vapour return line, but it can result in the tanker being completely full of liquid.

If it warms up, the pressure inside will rise. This does not matter if the tanker is fitted with a relief valve, but in the UK tankers containing toxic gases are not normally fitted with relief valves. On the Continent many tankers carrying liquefied flammable gases are also not fitted with relief valves[7]. One such tanker was overfilled in Spain in 1978 and burst, killing over 200 people[1,2]. (The tanker is believed to have been weakened by use with ammonia.)

Failure to keep the vapour return line open can also result in the overpressuring of tankers which do not have relief valves. After the insulation fell off a rail tanker while it was being filled, it was found that a valve in the vapour return line was closed and the gauge pressure was 23 bar instead of the usual 10 bar.

Relief valves are fitted to all pressure tankers in the US but they are not fitted in continental Europe. This is justified on the grounds that:

- frequent small leaks are more hazardous than an occasional burst, especially on rail tankers which have to pass through long tunnels;
- the valves may be knocked off in an accident;
- they are difficult to maintain on vehicles that are away from base for a long period;
- the relief valve is of no value if the vehicle falls over (as then the valve will discharge liquid);
- if a tanker is overpressured the manhole bolts will stretch and relieve the pressure.

In a typical UK compromise, relief valves are fitted to tankers carrying liquefied flammable gases but not those carrying liquefied toxic gases. Overfilling is prevented by following procedures and it is said that since 1945 no tanker has burst.

A cheap hardware solution to a problem, if there is one, is better than relying on procedures that, for one reason or another, may not always be followed. Small leaks from relief valves, the reason most often quoted for not using them, can be prevented by fitting bursting discs below the relief valves. Maintenance of the relief valves should be a small problem compared with maintenance of the vehicle as a whole. The case for and against relief valves should be re-examined[7].

11.2.3 A serious incident[3]

Automatic equipment was installed for loading tankers at a large oil storage depot. The tanker drivers set the quantity required on a meter, inserted a card which was their authorization to withdraw product, and then pressed the start button.

There was a manual valve in each filling line for use when the automatic equipment was out of order. To use the manual valves the automatic valves had first to be opened; this was done by operating a series of switches in the control room. They were inside a locked cupboard and a notice on the door reminded the operators that before operating the switches they should first check that the manual valves were closed.

The automatic equipment broke down and the supervisor decided to change over to manual filling. He asked the drivers to check that the manual valves were closed and then operated the switches to open the automatic valves. Some of the manual valves were not closed; petrol and other oils came out of the filling arms and either overfilled tankers or splashed directly on the ground. The petrol caught fire, killing three men, injuring 11 and destroying the whole row of 18 filling points.

It is easy to say that the accident was due to the errors of the drivers who did not check that the manual valves were closed or to the error of the supervisor who relied on the drivers instead of checking himself, but the design contained too many opportunities for error. In particular, the filling points were not visible to the person operating the over-ride switches.

In addition, the official report made several other criticisms:

- The employees had not been adequately trained. When training sessions were arranged no-one turned up as they could not be spared.
- The drivers had very little understanding of the properties of the materials handled.

- Instructions were in no sort of order. Bundles of unsorted documents were handed to the inspector for study.
- There were no regular inspections of the safety equipment.

A combination of error-prone hardware and poor software made an accident inevitable in the long run. To quote from the official report, '… had the same imagination and the same zeal been displayed in matters of safety as was applied to sophistication of equipment and efficient utilisation of plant and men, the accident need not have occurred.'

11.2.4 Loads delivered to the wrong place

A relief tanker driver arrived at a water treatment works with a load of aluminium sulphate solution. He was told where to deliver the load and given the key to the inlet cover on the tank. He discharged the load into the wrong tank, which the key also fitted. The solution entered the water supply and many customers complained of an unpleasant taste and, later, of diarrhoea, sore throats and green hair!

Earlier in the day, and soon after the tanker arrived, there had been blockages in the lime dosing pump. The operators therefore assumed that the complaints about taste were due to abnormal lime dosing and flushed out the lime dosing system. The real cause of the problem was not found until two days later[9].

When we have a problem and know there is something wrong, we assume it is the cause of the problem and stop thinking about other possible causes (see Section 4.4, page 88).

11.3 Some incidents that could be prevented by better instructions

This section is concerned with day-to-day instructions rather than the permanent instructions discussed in Section 3.5 (page 65). Sometimes care is taken over the permanent instructions but 'anything will do' for the daily instructions.

A day foreman left instructions for the night shift to clean a reactor. He wrote *Agitate with 150 litres nitric acid solution for 4 hours at 80°C*. He did not tell them to fill the reactor with water first. He thought this was obvious as the reactor had been cleaned this way in the past.

The night shift did not fill the reactor with water. They added the nitric acid to the empty reactor via the normal filling pump and line which contained some isopropanol. The nitric acid displaced the isopropanol into the reactor, and reacted violently with it, producing nitrous fumes. The reactor, designed for a gauge pressure of 3 bar, burst. If it had not burst, the gauge pressure would have reached 30 bar.

This accident could be said to be due to the failure of the night shift to understand their instructions or use their knowledge of chemistry (if any). It can be prevented only by training people to write clear, unambiguous instructions. No relief system can be designed to cope with unforeseen reactions involving chemicals not normally used.

A vented overhead tank was overfilled and some men were asked to clean up the spillage. They were working immediately below the tank which was filled to the top of the vent. A slight change in pressure in one of the lines connected to the tank caused it to overflow again — onto one of the men.

The plant where this occurred paid great attention to safety precautions when issuing permits for maintenance work and would have resented any suggestion that their standards were sloppy, but no-one realized that cleaning up spillages should receive as much consideration as conventional maintenance.

11.4 Some incidents involving hoses

Hoses, like bellows (see Section 8.1, page 155), are items of equipment that are easily damaged or misused, as the following incidents show, and while better training, instructions, inspections, etc., may reduce the number of incidents, we should try to change the work situation by using hoses as little as possible, especially for handling hazardous materials.

To extinguish a small fire a man picked up a hose already attached to the plant. Unfortunately it was connected to a methanol line. On another occasion a man used a hose attached to a caustic soda line to wash mud off his boots. In these cases a hardware solution is possible. If hoses must be left attached to process lines, then the unconnected ends should be fitted with self-sealing couplings.

A hose was secured by a screw coupling but only two threads were engaged. When pressure was applied the hose came undone with such force that the end of it hit a man and killed him.

Carelessness on the part of the man who connected the hose? A moment's aberration? Lack of training so that he did not realize the importance of engaging all the threads? It is safer to use flanged hoses if the pressure is high.

Hoses often burst or leak for a variety of reasons, usually because the wrong sort was used or it was in poor condition. People are often blamed for using the wrong hose or using a damaged hose. The chance of failure can be reduced by using the hose at as low a pressure as possible. For example, when unloading tankers it is better to use a fixed pump instead of the tanker's pump, as then the hose is exposed to the suction rather than the delivery pressure of the pump.

Why are hoses damaged so often? Perhaps they would suffer less if we provided hangers for them instead of leaving them lying on the ground to be run over by vehicles.

Accidents frequently occur when disconnecting hoses because there is no way of relieving the pressure inside and men are injured by the contents of the hose or by mechanical movement of it. For example, three men were sprayed with sediment when disconnecting a clogged hose at a quick-disconnect coupling. There was no way of releasing the pressure trapped between the positive pump and the blockage. In another similar incident five men standing up to ten feet away were sprayed with contaminated sludge[12].

The hardware solution is obvious. Fit a vent valve to the point on the plant to which the hose is connected. After an incident there is usually a campaign to fit such vents, but after two years hoses are connected to other points on the plant and the accident recurs. It is easy to blame the operators for not pointing out the absence of vent points. Hose connections should be inspected regularly.

There is a need for a design of vent valve that can be incorporated in a hose as standard. It should not project or it will be knocked off and it should not be possible to leave it open.

A man was sprayed with nitric acid mist when the unsecured end of a hose lifted out of a floor drain. The hose moved when the flow of liquid was followed by a rush of compressed air[13]. If the end of a hose is not connected to equipment, it should be secured.

11.5 Communication failures

These can of course affect design, construction and maintenance, but seem particularly common in operations, so they are discussed here.

11.5.1 Failure of written communication

Instructions are considered in Sections 3.5 and 11.3 (pages 65 and 192). Figure 11.2 shows three ambiguous notices.

Notice (a) appeared on a road tanker. The manager asked the filler why he had not earthed the tanker before filling it with flammable liquid. The filler pointed out the notice. It actually referred to the electrical system. Instead of

Figure 11.2

using the chassis as the earth, there was a wired return to the battery. It had nothing to do with the method used for filling the tanker.

Notice (b) is error-prone. If the full stop is not noticed, the meaning is reversed.

What does (c) mean?

Designers often recommend that equipment is 'checked' or 'inspected' but such words mean little. The designer should say *how often* the equipment should be checked or inspected, what should be looked for and *what standard* is acceptable.

The following appeared in a medical journal[10]:

MANAGEMENT OF SERIOUS PARACETAMOL POISONING
(12 December 1988)
Under the section on encephalopathy we said the patient should be nursed at 30–40 degrees. This referred to the angle in bed — not the temperature.

Pictorial symbols are often used in the hope that they will be understood by people who do not understand English and may not know the meaning of, for example, 'fragile'. However, pictorial symbols can be ambiguous. A storeman saw some boxes marked with broken wineglasses, meaning that the contents were fragile. He took the picture to mean that the contents were already broken and that therefore it did not matter how they were treated. The same man, who lived in a hot country, found some boxes marked with an umbrella, to indicate that they should not be allowed to get wet. He took the picture to mean that the boxes should be kept out of the sun.

An operator was found inside a confined space — the inside of an incinerator — although no entry permit had been issued. An industrial hygienist had written on the work permit that entry was acceptable. The operator took this to mean that entry had been approved but all the hygienist meant was that the atmosphere inside was safe to breathe. He did not mean that the isolation and other necessary precautions were in place. It was not his responsibility to authorize entry.

The following was printed on the base of a packet of food from a supermarket[14]:

Do not turn upside down

Two wartime newspaper headlines have become classic examples of ambiguity:

BRITISH PUSH BOTTLES UP GERMANS

MACARTHUR FLIES BACK TO FRONT

Two later examples are:

THREE MEN HELD IN CIGARETTE CASE

TRUMAN SEEKS MORE DAM CASH

"I DON'T KNOW HOW YOU CAN BE SO CARELESS, WAYNE—
THE CAN WAS CLEARLY MARKED TRINITROTOLUENE!"

11.5.2 Failures of verbal communication

The incidents described below were really the result of sloppy methods of working — that is, of poor management. People should not be expected to rely on word of mouth when mishearing or misunderstanding can have serious results. There should be better methods of communication.

For example, a famous accident occurred on the railways in 1873. Two trains were ready to depart from a station. The signalman called out, 'Right away, Dick,' to the guard of one train. Unknown to him, the guard of the other train was also called Dick[4].

A flat lorry was backed up against a loading platform and loaded with pallets by a fork-lift truck which ran onto the back of the lorry. When the fork-lift truck driver had finished, he sounded his horn as a signal to the lorry driver to move off. One day the lorry driver heard another horn and drove off just as the fork-lift truck was being driven off the lorry; it fell to the ground.

After a number of people had been scalded by hot condensate, used for clearing choked lines, it was realized that some operators did not know that 'hot condensate' was boiling water.

A member of a project team was asked to order the initial stocks of raw material for a new plant. One of them was TEA. He had previously worked on a plant on which TEA meant tri-ethylamine, so he ordered some drums of this chemical. Actually tri-ethanolamine was wanted. The plant manager ordered some for further use and the mistake was discovered by an alert storeman who noticed the two similar names on the drums and asked if both chemicals were needed.

A fitter was asked, by his supervisor, to dismantle heat exchanger 347C. The fitter thought the supervisor said 347B and started to dismantle it. The fitter should, of course, have been shown the permit-to-work, but it is easy to hear a letter incorrectly, particularly on the telephone. It is a good idea to use the international phonetic alphabet shown below. The heat exchanger would have been 347 Bravo, unlikely to be confused with 347 Charlie.

A	ALFA	J	JULIET	S	SIERRA
B	BRAVO	K	KILO	T	TANGO
C	CHARLIE	L	LIMA	U	UNIFORM
D	DELTA	M	MIKE	V	VICTOR
E	ECHO	N	NOVEMBER	W	WHISKEY
F	FOXTROT	O	OSCAR	X	X-RAY
G	GOLF	P	PAPA	Y	YANKEE
H	HOTEL	Q	QUEBEC	Z	ZULU
I	INDIA	R	ROMEO		

If there are two pumps on the same duty (one working, one spare) they are often labelled — for example, J25A and J25B. On one plant they were called instead J25 and JA25. Say the names out loud: Jay 25 and Jayay 25 sound too much alike. An electrician asked to replace the fuses in JA25 replaced them in J25 instead.

A contractor was employed to remove the surface coating from a floor. He was told to order some muriatic acid (that is, hydrochloric acid). He misheard and ordered 'muriatic acetone'[15]. There being no such substance, the supplier supplied acetone. It was ignited by a floor buffing machine. Two men were badly burned and several were exposed to high concentrations of acetone vapour.

Communication failures easily occur when valves (or other equipment) can be operated by more than one group of people. On one plant, when team A opened or closed valves they informed team B, who in turn informed team C. On one occasion team A closed an isolation valve for maintenance. No-one from team B was available so team A informed team C directly. Team B then tried to move some contaminated water through the closed valve to another part of the plant. The water backed up another line, flooded a pump room and seeped into several other rooms[16]. Only one team should be allowed to operate each item of equipment.

Operators often communicate with the control room by radio. The radios are often noisy and other people talking in the control room make communication difficult. In one case a field operator calmly reported that a tank was leaking and oil was filling the bund. The control room operator could not understand the message but, from the tone of the field operator's voice and the length of the message, concluded that everything was under control. When the level in the tank was seen to fall a supervisor went to investigate and found the operator desperately trying to contain the leak.

Nimmo has often asked control room operators who have been speaking on the radio, 'What did he say?' and got the answer that they did not know. They deduce the answer from the length of the sentences, key words and the tone of voice[17]. We should remember this when investigating incidents.

11.6 Examples from the railways

(a) Train doors

Many accidents have occurred because passengers opened train doors before the trains had stopped. For years, passengers were exhorted not to do so. A more effective method of prevention is automatic doors controlled by the crew, used on some suburban electric lines for many years, or doors which are automatically locked when the train is moving. In 1983 British Rail decided that these should be fitted to all new coaches[5].

(b) Off-loading tankers

While petrol was being offloaded from a train of rail tankers, they started to move, the hoses ruptured and 90 tons of petrol were spilt. Fortunately it did not ignite.

The tankers moved because the brakes were not fully applied. They could be applied in two ways:

- Automatically — the brakes were then held on by the pressure in compressed air cylinders. As the pressure leaked out of the cylinders, the brakes were released. This took from 15 minutes to over 4 hours, depending on the condition of the equipment.
- By hand — using a long lever which was held in position by a locking pin.

If the hand brake was applied while the brakes were already held on by compressed air, then when the air pressure fell the brakes were applied more firmly, were hard to release by hand and may have had to be dismantled. Instructions therefore stated that before the hand brakes were applied the compressed air in the cylinders must be blown off.

However, the operators found an easier way: they applied the hand brakes loosely so that when the air pressure leaked out the brakes would be applied normally. Unfortunately, on the day of the accident they did not move the hand brake levers far enough.

The design of the braking system was poor. Equipment should be designed so that correct operation is no more difficult than incorrect operation. The official report[6] did not make this point but it did draw attention to several other unsatisfactory features:

- The siding should be level.
- The hoses should be fitted with breakaway couplings which will seal if the hoses break.
- Emergency isolation valves or non-return valves should be fitted between the hoses and the collecting header.
- The supervisors should have noticed that brakes were not being applied correctly.

(c) Single-line working

Two freight trains collided head-on on a single-line railway in the United States, killing three crew members and seriously injuring a fourth. Six locomotives were destroyed and the total damage was $4 million. There was no signalling system, no automatic equipment to prevent two trains entering the same section of track and not even a single-line token system such as that used in the UK (in which the driver has to be in possession of a token for the section of line). A 'dispatcher' authorized the two trains to enter the line in opposite directions and the official report[11] gave the cause of the accident as 'inadequate personnel selection criteria which resulted in the placement of an individual without sufficient training or supervision into the safety critical position of train dispatcher'. While this may have contributed to the accident, even an experienced man could have had a moment's lapse of attention and either additional safety equipment or a better method of working should have been recommended.

11.7 Simple causes in high tech industries

Very often, even in high technology industries, the causes of failures and accidents are elementary. Some are described in Section 9.1 (page 168). Here are a few more:

- A pump bearing failure was traced to poor lubrication. Drums of oil had been left standing out-of-doors with the lids off, allowing rain to enter.
- Some welds in purchased equipment showed unexpected failures. The materials experts from the purchasing company visited the supplier and discussed various possible but esoteric reasons for the failures. They then asked if they could see the workshop. They found that welding was carried out near a large open door, protected only by plastic sheets, and that dust was blown about by the wind.
- The cooling coils in a storage vessel showed unexpected corrosion. I attended a meeting at which possible causes were discussed. A materials expert outlined several possible but complex chains of events. I then asked if the pipework had been checked before installation to make sure that it was made from the grade of steel specified. It had not been checked (see Section 9.1, penultimate paragraph, page 170).
- The nuclear industry has a good safety record. One of the few accidents at one nuclear site occurred when a ceiling tile fell 20 feet and injured a man in the canteen. A steam trap in the space above the ceiling had been leaking. It was replaced but no-one checked that the repair was successful. It continued to leak and water accumulated[18]. The following two incidents also occurred at nuclear sites.
- As reported in Section 3.4 (page 62), an aqueous cleaning solution was sprayed onto electrical equipment, causing arcing and burning a 6-inch hole in the casing.

- Hazardous laboratory wastes were placed in a plastic bottle for disposal. No-one kept a record of what was in it, different wastes reacted and the bottle exploded[19].

- On several occasions cylinders of hydrogen fluoride (HF) have ruptured after being kept for 15–25 years. The HF reacts with the steel to form iron fluoride and hydrogen[20].

References in Chapter 11

1. Hymes, I., 1983, *The Physiological and Pathological Effects of Thermal Radiation, Report No. SRD R 275* (UK Atomic Energy Authority, Warrington, UK).

2. Stinton, H.G., 1983, *Journal of Hazardous Materials*, 7: 393. (This report contains a number of errors.)

3. *Official Report on the Fire at the West London Terminal of Esso Petroleum*, 1968 (HMSO, London, UK).

4. Steele, A.K., 1972, *Great Western Broad Gauge Album* (Oxford Publishing Company, Oxford, UK).

5. *Modern Railways*, 1985, 42(1): 42.

6. Health and Safety Executive, 1984, *Report on a Petroleum Spillage at Micheldover Oil Terminal Hampshire on 2 February 1983* (HMSO, London, UK).

7. Kletz, T.A., 1986, *Plant/Operations Progress*, 5(3): 160.

8. Hakkinen, K., 1983, *Scandinavian Journal of Work and Environmental Health*, 9: 189.

9. *Chemistry in Britain*, 1988, 24(10): 980.

10. *Drug and Therapeutics Bulletin*, quoted in *Atom*, 1990, No. 400, page 38.

11. *News Digest*, 1985, 4(2) (National Transportation Safety Board, Washington, DC, USA).

12. *Operating Experience Weekly Summary*, 1998, No. 98–47, page 8 (Office of Nuclear and Facility Safety, US Department of Energy, Washington, DC, USA).

13. *Operating Experience Weekly Summary*, 1998, No. 98–17, page 6 (Office of Nuclear and Facility Safety, US Department of Energy, Washington, DC, USA).

14. Armstrong, S., *The Times*, 1 December 1998.

15. *Operating Experience Weekly Summary*, 1998, No. 98–51, page 10 (Office of Nuclear and Facility Safety, US Department of Energy, Washington, DC, USA).

16. *Operating Experience Weekly Summary*, 1998, No. 98–41, page 2 (Office of Nuclear and Facility Safety, US Department of Energy, Washington, DC, USA).

17. Nimmo, I., 1995, *Chemical Engineering Progress*, 91(9): 36.

18. *Operating Experience Weekly Summary*, 1998, No. 99–06, page 14 (Office of Nuclear and Facility Safety, US Department of Energy, Washington, DC, USA).

19. *Operating Experience Weekly Summary*, 1998, No. 98–18, page 1 (Office of Nuclear and Facility Safety, US Department of Energy, Washington, DC, USA).

20. *Operating Experience Weekly Summary*, 1999, No. 99–25, page 12 (Office of Nuclear and Facility Safety, US Department of Energy, Washington, DC, USA).

Errors in computer-controlled plants

12

'To err is human. To really foul things up needs a computer.'
Anon

'Computers allow us to make more mistakes, faster than ever before.'
Anon

The uses of computer control continue to grow and if we can learn from the incidents that have occurred we may be able to prevent repetitions. The failures are really human failures: failures to realize how people will respond; failures to allow for foreseeable faults.

The equipment used is variously described as microprocessors, computers and programmable electronic systems (PES). The last phrase is the most precise as microprocessors do not contain all the features of a general purpose digital computer. Nevertheless I have used 'computer' throughout this chapter because it is the word normally used by the non-expert.

The incidents described are classified as follows:

12.1 Hardware failures: the equipment did not perform as expected and the results of failure were not foreseen (see Figure 12.1).

12.2 Software errors: errors in the instructions given to the computer.

12.3 Specification errors, including failures to understand what the computer can and cannot do. This is probably the most common cause of incidents.

12.4 Misjudging the way operators will respond to the computer.

12.5 Errors in the data entered in the computer.

12.6 Failure to tell the operators of changes in data or programs.

12.7 Unauthorized interference with the hardware or software.

For more information on the subject of this chapter see References 1, 17, 18, 19 and 34. References 19 and 34 are thorough examinations of the subject.

Figure 12.1

12.1 Hardware failures

In many cases the results of failure could reasonably have been foreseen and precautions taken or designs changed, as shown by the examples below.

It is generally agreed that the ultimate safety protection of plant or equipment should be independent of the control system and should be hardwired or based on an independent computer system. Alarms can be part of the control system, but not trips and emergency shutdown systems. Interlocks which prevent a valve being opened unless another is shut are an intermediate category. They are often part of the control system, but where the consequences of failure are serious, independence is desirable.

For example, a hardware failure caused a number of electrically-operated isolation and control valves to open at the wrong time. Hot polymer was discharged onto the floor of a building and nitrogen, used for blanketing the vessels from which the spillage occurred, was also released. A watchdog which should have given warning of the hardware failure was affected by the fault and failed to respond[2,3]. On the plant concerned, trip systems were independent of the control computer but the designers did not realize that other safety systems, such as interlocks to prevent valves being open at the wrong time, should also be independent. Also, watchdogs should not be affected by failures elsewhere. A hazard and operability study which questioned what would happen if foreseeable hardware failures occurred, at each stage of a batch, would have disclosed the design faults.

Note that the initial hardware failure was merely a triggering event. It would not have had serious results if the watchdog or the interlocks had been truly independent. The underlying cause was a failure to separate the control and the safety systems.

The vendors of a microprocessor-based fire protection system have reported that power transients and nearby electric fields can produce false alarms[20].

An incident from another industry: on 3 June 1980 the screens at US Strategic Air Command showed that missiles were heading towards the US. The immediate cause of the false alarm was a hardware failure. The system was tested by sending alarm messages in which the number of missiles was shown as zero. When the hardware failure occurred the system replaced zero by random numbers. The software had not been designed so as to minimize the effects of hardware failure[4,5].

12.2 Software errors

These can be subdivided into errors in the systems software, bought with the computer, and errors in the applications software, written for the particular application. To reduce the first, if possible only well-tested systems should be used — not always easy in a field that is changing rapidly. To reduce the latter, thorough testing is essential. It can take longer than design.

Software errors are similar to the slips discussed in Chapters 2 and 7. Gondran[6] quotes the following figures for the probability that there will be a significant error in the applications software of a typical microprocessor-based control system:

Normal system:	10^{-2}–10^{-3}
Considerable extra effort:	10^{-4}
Additional effort is as great as the initial development effort:	10^{-6}

Others consider that the figure for normal systems is too optimistic for all but the simplest systems.

The number of possible paths through a system can be enormous and it may be impossible to test them all. When we have found that an error is present it is often difficult to locate the fault. Wray[7] writes, 'I was involved with a machine which failed to stop when a man put his hand through a photo-electric guard; fortunately he wasn't injured but I've now got a program of 279 pages of assembly code with the comments in German and which I suspect contains a fault.' (It was found to be easier to scrap the software and start again.)

Leveson[19] recommends that instead of trying to make software ultra-reliable, we should try to make the plant safe even if the software is unreliable. This can be done in two ways: by installing independent trips and interlocks, the defence-in-depth approach, or, preferably, by developing when possible user-friendly and inherently safer designs — that is, plant and equipment designs in which the hazards are avoided rather than kept under control, as discussed in Sections 6.7.1 and 8.7 (pages 120 and 162).

Of course, we should design such plants whenever we can, regardless of the type of control, and we should be on our guard against the view, sometimes heard but more often implicit, 'Don't worry about the hazards; the control system will keep them under control.' This is not a view held by control engineers, who know only too well the limitations of their equipment and the ways in which it can be neglected, but other engineers sometimes hold this view.

Modifications to software can introduce faults. Software changes (and hardware changes (see Section 12.7.1, page 214) should be treated as seriously as changes to plant or process and subject to similar control. No change should be made until it has been authorized by a responsible person who should first carry out a systematic study of the consequences by Hazop or similar techniques. Access to software should be restricted.

While hardware errors are usually random (as equipment is rarely in use long enough for wear-out failures to occur), software failures are more like time bombs. They lie in wait until a particular combination of process conditions occurs.

In the first incident described in Section 12.1 (page 204), the software had been extensively modified and was very different from that originally installed. However, the errors were of the type discussed in Section 12.3 (page 207).

Software errors are rather like spelling and grammatical errors in written instructions. However, people usually know what is meant despite spelling or grammatical errors. We understand what we are expected to do if we are told to 'Save soap and waste paper' or to 'Wash the teapot and stand upside down in the sink'. Computers can do only what they are told to do.

However, even though our spelling and grammar are correct our instructions, to people or computers, may still be wrong. This is discussed in the next section.

12.3 Specification errors

A number of incidents have occurred because a computer failed to perform in the way that the designer or operators expected it to perform. These incidents did not occur because of equipment faults or errors in the software, but because the program logic was not specified correctly. Sometimes the software engineer did not understand the designer's or operators' requirements or was not given sufficiently detailed instructions covering all eventualities. Sometimes the operators did not understand what the computer could and could not do. They treated it as a black box, something that will do what we want even though we do not understand what goes on inside it. The following incidents are typical of many.

The first occurred on a rather simple batch reactor control system (Figure 12.2, page 208). The computer was programmed so that, when a fault occurred on the plant, it would sound an alarm and then hold everything steady until the operator told it to carry on.

One day the computer received a signal that there was a low oil level in a gearbox. It sounded the alarm and kept all its output signals steady. The computer had just added the catalyst to the reactor and it would normally have increased the cooling water flow to the condenser. It kept the flow steady, the reactor overheated and the contents were lost to atmosphere through the relief valve. The operator was busy investigating the cause of the alarm (it turned out to be a false alarm) and did not notice the rising temperature.

A Hazop had been carried out on the plant but had not included the computer. It had been treated as a black box, something that will do what we want without the need to understand how it works or what are its limitations. During the Hazop the team should have asked what actions the computer would take, for all possible deviations, at all stages of the batch. Actions which would affect safety or operability would then have come to light. The software engineer should have been a member of the Hazop team.

Blanket instructions are undesirable whether directed at people or computers. We should always check them by considering in turn all possible

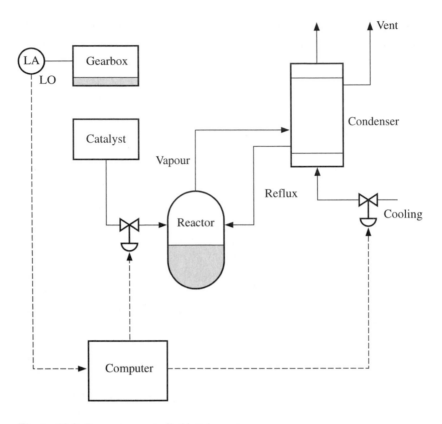

Figure 12.2 Computer-controlled batch reactor

situations. This would probably have been done if the instructions had been addressed to the operators, but the design team acted as if the computer would somehow cope with any problems.

Note that an operator, if given the same instructions as the computer, would have used his discretion and increased the cooling before investigating the reason for the alarm. Men can do what we want them to do but a computer can do only what we tell it to do. To use the expressions used in earlier chapters, computers can carry out only rule-based actions, not knowledge-based actions. The rules can be more complex than a man could handle but will be followed to the letter when even the dimmest operator will see the need for an exception. (Expert systems convert what seem like knowledge-based decisions into rule-based ones.)

When the manager asked the software engineer to keep all variables steady when an alarm sounds, did he mean that the cooling water flow should be kept steady or that the reactor temperature should be kept steady? He probably never thought about it. A Hazop would have provided an opportunity to discuss this point.

One way of solving communication problems is to combine the jobs of those who find communication difficult. There is a need for technologists who are equally at home in the fields of process engineering and computer control. There are such people, greatly valued by their managers, but we need more and I know of no courses that set out to train them.

Another incident occurred on a pressure filter controlled by a computer. To measure the pressure drop through the cake, the computer counted the number of times the air pressure in the filter had to be topped up in 15 minutes. If less than five top-ups were needed, filtration was complete and the computer could move on to the next phase, smoothing the cake. If more than five top-ups were needed, the liquid was circulated for a further two hours.

There was a leak of compressed air into the filter, which misled the computer into thinking that filtration was complete. It signalled this fact to the operator, who opened the filter door and the entire batch — liquid and solid — was spilt.

To be fair to the computer, or rather to the programmer, the computer had detected that something was wrong — there was no increase in power consumption during smoothing — and had signalled this fact by stopping the operation, but the operator ignored this warning sign, or did not appreciate its significance[8].

Again a Hazop would probably have disclosed the weakness in the system for detecting the pressure drop through the cake and changes could have been made. In particular it would be desirable to provide some means of preventing the operator opening up the filter while it is full of liquid. Many accidents have occurred because operators opened up autoclaves or other pressure vessels while they were up to pressure (see Section 2.2.1, page 13). Opening up a vessel while it is full of liquid at low pressure is not as dangerous but nevertheless dangerous enough. Chapter 2 of Reference 1 discusses Chazops, variations of Hazop suitable for computerized systems.

As these incidents show, we should not treat the computer as a black box, but should understand what it will do, under all foreseeable conditions. It is not necessary to understand the electronics but we should understand the logic. In place of black boxes we need 'glass boxes' in which the logic is visible to all.

One of the problems is that the software is incomprehensible to most of the customers for whom it is written. If a process engineer asks someone to write a conventional instruction for him, he will read it through to check that his intentions have been understood and carried out. If a process engineer asks for a certain method of control, he will look at the process and instrumentation diagram to check that his intentions have been understood and carried out. Every process engineer can read a process and instrumentation drawing; the 'language' is soon learnt. In contrast, if a process engineer asks a software engineer to prepare a computer program, the result is no more readable than if it was written in Babylonian cuneiform. Computer languages are difficult or impossible to read unless one is thoroughly familiar with them and using them frequently; they are not scrutable and cannot readily be checked. Even the specification — the instructions to the programmer — may not be easy to check.

If we want to see that something will do what we want, we normally look at the final product, not the instructions for making it. I want to see food in the supermarket before I buy it, not recipes. In the long term perhaps we shall see the development of plain English languages which anyone can read[21]. Until this time comes we shall have to make do with the specifications, but at least we should subject them to a thorough Hazop.

To sum up this section, here is a quotation from an HSE report[18]:

'The achievement of the required reliability by the hardware and software alone, however, is not enough to guarantee overall plant safety. If the specification of the safety-critical system is inadequate then the overall nuclear system may be unsafe even though the hardware and software implementation of the safety-system is completely reliable (with respect to its specification).'

12.4 Misjudging responses to a computer

Adding alarms to a computerized control system is cheap and easy. The operators like to know when a measurement changes on a page that is not on display at the time so that they can intervene if necessary, so more and more alarms are added. When a serious upset occurs hundreds of alarms may sound at once, the operators have no idea what is happening and they switch off (themselves, not the computer). In addition, low priority alarms can divert operators' attention from more important ones. Fitting alarms illustrates the maxim that if we deal with each problem as it arises, the end result may be the opposite of that intended (see also Section 4.2.1, page 80).

As discussed in Section 4.2.3 (page 83), the operator should feel that he is in charge of the plant and not a passive bystander. To quote Reference 18 again:

'Human beings can only maintain the skills and commitments needed to carry out tasks by actually doing those tasks, and increased computer operation may give the operator less opportunity for this. So any increase in computer responsibilities means that more care is needed in supporting operators by interface and job design, and by training, to ensure that they can carry out their remaining tasks effectively.'

Another problem with alarms is that the information operators need for handling alarms is often distributed amongst several pages of the display. It should be possible to link together on a special page for each significant alarm the information needed for dealing with it.

The pages of a display sometimes look alike. This saves development time, and thus cost, but can cause confusion in an emergency because an operator may turn to the wrong page and not realize he has done so.

Data are usually written onto a disk at intervals. These should be frequent. On one plant the computer collected spot values of each instrument reading at the beginning of every minute and then, every five minutes, it wrote them onto a hard disk. The hard disk survived a runaway reaction and an explosion but all other data were lost. The explosion occurred near the end of a five-minute period and nearly five minutes' data were therefore lost. The highest pressure recorded was 4 bar although the bursting pressure of the reactor that burst was about 60 bar[22].

The following incident occurred on a plant where the possibility of a leak of liquid had been foreseen and a sump had been provided into which any leaks would drain. A level alarm would then sound. Unfortunately, when a leak occurred it fell onto a hot pipe; most of it evaporated, leaving a solid residue and none entered the sump. The leak was not detected for several hours.

The operators could have detected that something was wrong by a careful comparison of trends in a number of measurements but they saw no need to make such a comparison because they were not aware of any problem. The relevant measurements were not normally displayed and had to be called up. Afterwards the operators said that the spillage would have been detected earlier if the chart recorders had not been removed from the control room when the computers were installed. The system could have been programmed to carry out mass balances, compare readings for consistency and/or to sound an alarm (or, better, display a message advising the operators that something was amiss) when unexpected measurements were received, but no-one had foreseen the need to ask it to do so. More simply, additional screens could be installed to continuously show trends in important parameters.

The limit switch on an automatic valve (A) in a feedline to a reactor was out of order. Whatever the position of the valve, the limit switch told the computer that the valve was open so the operator sent for an instrument mechanic.

When the mechanic arrived the valve was actually open. In order to check whether or not the limit switch was faulty, he sent a false signal to the computer to say that the limit switch was closed. The computer received the signal that the valve had closed and moved on to the next stage of the process — addition of hydrogen to the reactor. The hydrogen passed through valve A, in the wrong direction, into a low-pressure vessel, which ruptured.

The mechanic and the operators did not foresee what would happen when a false signal was sent to the computer[9].

As on plants which are not computer-controlled, we should ask, during Hazops, what will happen when foreseeable faults occur.

A batch plant had two parallel trains, A and B, which could be controlled (remote manually) from any of three control stations. It was custom and practice to connect control station 1 to stream A, control station 3 to stream B and control station 2 to either stream.

Stream B was shut down. A number of operators wanted to observe what was happening on stream A and so stations 1 and 2 were connected to it. An engineer came into the control room to check on progress on stream A. As there was a crowd around stations 1 and 2, he used station 3. He then left stream A on display.

Soon afterwards an operator arrived to prepare stream B for start-up. As station 3 was normally used only for stream B, he assumed it was the one on display and operated several valves. He actually moved the valves on stream A and ruined the batch[23].

There were, of course, errors by the engineer who failed to follow custom and practice and to switch off the control station when he left, and by the operator who did not check that station 3 was connected to stream B. But the system was user-unfriendly. Different colours could have been used to distinguish the A and B displays. Station 1 could have been permanently connected to A, and station 3 to B, while 2 could have been a read-only station connected to either. As in one of the incidents described in Section 11.6 (page 199), only the operators responsible for A should be allowed to operate valves on A or make other changes.

Accidentally hitting the wrong button should not produce serious effects but there have been some such incidents in the financial markets. For example, a bank trader in London accidentally and repeatedly hit an 'instant sell' button; 145 sell orders for French 12-year bonds caused their price to fall drastically[24].

Another incident occurred when a clerk at the Australian Reserve Bank was asked to send an e-mail message announcing an increase in interest rates from 5 to 5.5% with effect from 9.30 am. She typed the message at 9.24 am, and specified a delay of 6 minutes before transmission. However, she forgot that a separate menu had to be called up to activate the delay. As a result the computer carried out its default action and sent the message immediately. At 9.24 am the Australian dollar was worth US\$63.20. So much Australian currency was bought during the next five minutes that at 9.30 am the Australian dollar was worth US\$63.70. Speculators made a profit of four million Australian dollars[35,36].

The bank admitted that there were shortcomings in their procedures but nevertheless the clerk and her supervisors were transferred to other positions and their salaries were reduced, an action about as effective in preventing further incidents as announcing after someone has slipped on an icy road that no-one should do so in future. If any people were responsible they were the senior managers who did not have their systems reviewed for features likely to produce errors.

To quote John Humphrys, 'Responsibility has been redefined to include *being seen to do something*, anything, in response to transient public sentiment often generated by television images, rather than coolly assessing what are realistically the best options'[37].

12.5 Entering the wrong data

An operator calculated the quantities of reagents required for a batch reaction and asked the foreman to check his figures. The foreman found an error. By this time a shift change had occurred, the new operator did not realize that one of the figures had been changed and he used the original ones. The error was not hazardous but the batch was spoilt.

During the Hazop of computer-controlled plants we should consider the results of errors in entering data. Accidental entering of wrong data can be minimized by requiring operators to carry out at least two operations — for example, moving a cursor and entering figures and, if necessary, the computer can be programmed to reject or query instructions which are outside specified ranges or do not satisfy consistency tests.

Entering the wrong data has caused incidents in other industries and activities. A pilot set the heading in a plane's inertial navigation system as 270° instead of 027°. The plane ran out of fuel and had to land in the Brazilian jungle. Twelve people were killed[10].

A patient called Simpkin was admitted to hospital with suspected meningitis. A first test was negative and she was diagnosed as suffering from acute influenza. A second test was positive but the result was entered into the computer under the name Simkin and was thus never seen by her doctor. The patient died[25]. Like the incidents discussed in Chapter 2, it is far too simplistic to say that the death was due to a slip by the person who entered the data. The program should have contained safeguards such as the need to check that there was a patient with that name in the hospital and that she was suffering from suspected meningitis.

A lady had an emergency pacemaker fitted on 4 November 1996 while visiting Canada. She was told to arrange for it to be checked a month later, when she had returned to the UK. On asking for an appointment at her local hospital she was surprised to be given one for April 1997. The date of the operation was shown on her Canadian notes as 11/4/96. The UK hospital interpreted this as 11 April 1996 and gave her an appointment for an annual check[26]. Now that so much communication is international, months should be spelt, to avoid ambiguity.

12.6 Failures to tell operators of changes in data or programs

I do not know of any incidents in the process industries that have occurred for this reason but it was the cause of an aircraft crash. In 1979 an Air New Zealand plane flew into a mountainside, with the loss of 257 lives, while on a sight-seeing tour of Antarctica. Unknown to the crew, the co-ordinates of the destination way-point had been moved 2 degrees to the East. The inertial navigation system guided the plane, which was flying at low altitude so that the passengers could view the scenery, along a valley that ended in a cliff. It looked very similar to the valley that the crew expected to follow and they did not realize that they were on the wrong course[11,12].

12.7 Unauthorized interference with hardware or software

12.7.1 Hardware

A plant could be operated manually or by computer control. It was switched to manual control to carry out some maintenance on a reactor. One of the valves connected to the reactor had to be closed but an interlock prevented it closing.

The connections to the limit switches on the valve were therefore interchanged so that the valve was closed when the computer thought it was open.

When the maintenance was complete, the plant was switched back to computer control before the limit switches were restored to their normal positions. The computer thought the valve was open and decided to close it. It actually opened it, releasing flammable material[2].

This incident, like most of those described in this chapter, was not really the result of using a computer. It was the result of a totally unsatisfactory method of preparing equipment for maintenance, a frequent cause of accidents on plants of all sorts (see Chapter 10). The incident could have occurred on manual control if the operator had forgotten that the switches were interchanged. Perhaps he had forgotten. Or perhaps he expected the all-powerful computer to somehow know what had happened. According to the report on one incident, even when alarms were sounding, the operator did not believe it was a real emergency; 'the computer can cope'. Eberts says that some operators expect computers to behave like humans and cannot understand why they make mistakes that no human would make[13].

Files had to be transferred from a control computer to a training simulator. At first, there was no direct connection; the files were first transferred to a free-standing workstation and then from there to the simulator. To simplify the transfer a direct cable connection was made between the control computer and the simulator.

Unfortunately the address of the gateway in the control computer used for the data transfer was the same as that used to connect to the distributed control system (dcs). As a result, data flowed from the simulator through the control computer to the dcs and replaced the current input data by historic data. Some conditions on the plant started to change. Fortunately this was soon noticed by alert operators and the plant brought back under control.

This incident can teach us several lessons:

- No modification should be made to hardware (or software; see Section 12.2, page 205) until a systematic attempt has been made to identify the consequences and it has been authorized by a professionally qualified person, such as the first level of professional management. Since the explosion at Flixborough in 1974 this principle has been widely accepted for modifications to process equipment and operating conditions; it should also be applied to modifications to computerized systems.
- Connecting a control computer to another system is a modification and should only be carried out after systematic study of possible consequences. If made, data flow should be possible only in the outward direction. The possible results of access by malevolent hackers are serious.

215

- All systems should be secure. Houses need doors. The doors on control systems are less tangible than those on houses but just as important.

12.7.2 Software

Suppose an operator has difficulty charging a reactor because the vacuum is too low. He assumes the vacuum transmitter is faulty and decides to override it. He is able to do so by typing in an acceptable vacuum measurement in place of the figure received from the transmitter. Sometimes the operator is right and production continues. On other occasions the transmitter is not faulty and an incident occurs[14].

It is usually more difficult today for operators to interfere with the software in this way than in the early days of computer control. However, some companies allow operators to have 'keys' which allow them to override data, change interlock settings and so on. Other operators acquire them in various ways, much as operators have always acquired various tools and adaptors that they were not supposed to have.

While we may never be able to prevent completely this sort of action by operators, we can at least make it difficult. And we should ask ourselves why operators find it necessary to acquire illegal tools and keys.

12.7.3 Viruses

I have seen only one report of a virus in computer control software. It was found in the central computer of a Lithuanian nuclear reactor and was said to have been inserted by the person in charge of the safety-control systems, so that he could help in the investigation and demonstrate his ability[27]. However, any disruption of computerized control systems could be more serious than loss of accounting or technical data[15]. Once programs have been set up, they are added to infrequently. For infection to occur, viruses would have to be present in the original software or introduced via connections to other computers. At one time, plant control computers were rarely connected to other computers or, if they were connected, the flow of data was outwards only, but inward flow is becoming increasingly common (as described in Section 12.7.1, page 214). For example, a production control computer may tell a process control computer that more (or less) output is wanted. Viruses can also be introduced by allowing operators to play computer games.

Computer viruses are rather like AIDS. Do not promiscuously share data and disks and you are unlikely to be affected[16].

This is not as simple as it might seem. If the machine in charge of a production process is a programmable controller, we would expect it to be immune to viruses or compatibility problems. A controller does not have a disk drive, and there is no operating system. However, when the time comes to upgrade the facility, the controller program has to be changed. The new program may be generated on a PC.

12.8 The hazards of old software

In the process industries the hazards of reusing old equipment are well documented, though this does not mean that incidents no longer occur. Old equipment may be built to lower standards than those used today, it may have been modified during its life, it may have suffered from exposure to extreme conditions of use and it may require so much modification for its new duty that it might be simpler and cheaper to scrap it and start again.

All these reservations apply to old software, except that (unfortunately?) it never wears out. Leveson[28] has described an incident on the Therac, an apparatus for irradiating cancer patients, Shen-Orr[29] has described other incidents and Lions[30] has described the loss of the Ariane 5 space rocket:

'In Ariane 4 flights using the same type of inertial reference system there has been no such failure because the trajectory during the first 40 seconds of flight is such that the particular variable related to horizontal velocity cannot reach, with an adequate operational margin, a value beyond the limit present in the software.

'Ariane 5 has a high initial acceleration and a trajectory which leads to a build-up of horizontal velocity which is five times more rapid than for Ariane 4. The higher horizontal velocity of Ariane 5 generated, within the 40-second timeframe, the excessive value which caused the inertial system computers to cease operation.'

A function that did not serve any purpose in Ariane 5 was thus left in 'for commonality reasons' and the decision to do so 'was not analysed or fully understood'.

12.9 Other applications of computers

If any calculations are carried out as part of a control or design program, their accuracy should be checked by independent calculation. For example, manual checks of a program for calculating pipe stresses showed that gravitational stresses had been left out in error[31].

Pertrowski gives the following words of caution[32]:

'... a greater danger lies in the growing use of microcomputers. Since these machines and a plethora of software for them are so readily available and so inexpensive, there is concern that engineers will take on jobs that are at best on the fringes of their expertise. And being inexperienced in an area, they are less likely to be critical of a computer-generated design that would make no sense to an older engineer who would have developed a feel for the structure through the many calculations he had performed on his slide rule.'

Of course, one does not need a computer to apply formulae uncritically outside their range of applicability. I have described examples elsewhere[33]. As with almost everything else in this chapter, computers provided new opportunities for familiar errors. We can make errors faster than before.

Some software contains codes which prevent it being used after a prescribed period if the bill has not been paid.

12.10 Conclusions

As we have seen, computers do not introduce new sorts of error, but they provide new opportunities for making old errors. On any plant incidents can occur if we do not allow for the effects of foreseeable errors or equipment failures, if operators are given blanket instructions or are not told of changes in control settings, operating instructions or batch compositions, if there is poor handover at shift change or if equipment or procedures are changed without authority. However, some of the incidents are particularly likely to occur on computer-controlled plants because different departments may be responsible for operation of the plant and changes to the computer program, operators (and designers) may have exaggerated views of the computer's powers and many people may have a limited understanding of what it can and cannot do and how it does it.

References in Chapter 12

1. Kletz, T.A., Chung, P.W.H., Broomfield, E. and Shen-Orr, C., 1995, *Computer Control and Human Error* (Institution of Chemical Engineers, Rugby, UK).
2. Eddershaw, B.W., 1989, *Loss Prevention Bulletin*, No. 088, page 3.
3. Nimmo, L., Nunns, S.R. and Eddershaw, B.W., 1987, Lessons learned from the failure of computer systems controlling a nylon polymer plant, *Safety & Reliability Society Symposium, Altrincham, UK, November 1987*.
4. Borning, A., 1987, *Communications of the Association for Computing*

Machinery, 30(2).

5. Sankaran, S., 1987, Applying computers to safe control and operation of hazardous process plants, *Instrument Asia 87 Conference, Singapore, May 1987*.

6. Gondran, M., 1986, *Launch meeting of the European Safety and Reliability Association, Brussels, October 1986*.

7. Wray, A.M., 1986, *Design for Safe Operation — Experimental Rig to Production*, page 36 (Institute of University Safety Officers, Bristol, UK).

8. *Quarterly Safety Summary*, 1985, 55(221): 6 (Chemical Industries Association, London, UK).

9. *Quarterly Safety Summary*, 1984, 55(220): 95 (Chemical Industries Association, London, UK).

10. Learmont, D., *Flight International*, 17–23 January 1990, page 42.

11. Mahon, P., 1984, *Verdict on Erebus* (Collins, Auckland, New Zealand).

12. Shadbolt, M., *Reader's Digest*, November 1984, page 164.

13. Eberts, R.E., 1985, *Chemical Engineering Progress*, 81(12): 30.

14. Gabbett, J.F., 1982, PVC computer control experience, *AIChE Loss Prevention Symposium, Anaheim, CA, 1982*.

15. Ehrenburger, W.D., 1988, *European Safety & Reliability Association Newsletter*, 5(2).

16. Becket, M., *The Daily Telegraph*, 26 September 1989.

17. Health and Safety Executive, 1995, *Out of Control: Why Systems go Wrong and How to Prevent Failure* (HSE Books, Sudbury, UK).

18. Health and Safety Executive, 1998, *The Use of Computers in Safety-critical Applications* (HSE Books, Sudbury, UK).

19. Leveson, N.G., 1995, *Safeware — System Safety and Computers* (Addison-Wesley, Reading, MA, USA).

20. *Operating Experience Weekly Summary*, 1999, No. 99–07, page 5 (Office of Nuclear and Facility Safety, US Department of Energy, Washington, DC, USA).

21. Siddall, E., 1993, *Computer Programming in English, Paper No. 33* (Institute for Risk Research, University of Waterloo, Canada).

22. Mooney, D.G., 1991, An overview of the Shell fluoroaromatics explosion, *Hazards XI — New Directions in Process Safety, Symposium Series No. 124* (Institution of Chemical Engineers, Rugby, UK).

23. Hendershot, D.C. and Keeports, G.L., 1999, *Process Safety Progress*, 18(2): 113.

24. *Financial Times*, 19 October 1998.

25. Fleet, M., *Daily Telegraph*, 28 August 1998.

26. Foord, A.G., *Safety Critical Systems Newsletter*, May 1997, page 7.

27. Sagan, S. in Dutton, W.H. *et al.* (editors), 1995, *Computer Control and Human Limits: Learning from IT and Telecommunications Disasters*, page 37 (Programme on Information and Communications Technologies, Brunel University, Uxbridge, UK).

28. Leveson, N., 1995, *Safeware — System Safety and Computers*, Appendix A (Addison-Wesley, Reading, MA, USA).

29. Shen-Orr, C. in Kletz, T.A. (editor), 1995, *Computer Control and Human Error*, Chapter 3 (Institution of Chemical Engineers, Rugby, UK).
30. Lions, J.L., 1996, *Ariane 5 — Flight 501 Failure* (European Space Agency, Paris, France).
31. *Errors in Commercial Software Increase Potential for Process Piping Failures, Bulletin No. 89–B*, 1989 (US Department of Energy, Washington, DC, USA).
32. Petrowski, H., 1982, *To Engineer is Human* (St Martin's Press, New York, USA).
33. Kletz, T.A., 1996, *Dispelling Chemical Engineering Myths*, page 123 (Taylor & Francis, Philadelphia, PA, USA).
34. Neumann, P.G., 1995, *Computer Related Risks* (ACM Press, New York, USA).
35. *The Australian*, 19 February 2000.
36. Tweeddale, H.M., 2000, Nourishing and poisoning a 'safety culture', *Chemeca Conference, Perth, Australia*.
37. Humphrys, J., 1999, *Devil's Advocate*, page 63 (Hutchison, London, UK).

Personal and managerial responsibility

13

'*Personal responsibility is a noble ideal, a
necessary individual aim, but it is no use basing
social expectations upon it, — they will prove to
be illusions.*'
B. Inglis, 1964, *Private Conscience —
Public Morality*, 138

'*There is almost no human action or decision that
cannot be made to look flawed and less sensible
in the misleading light of hindsight.*'
Report on the Clapham Junction railway accident[16]

13.1 Personal responsibility

The reader who has got this far may wonder what has happened to the old-fashioned virtue of personal responsibility. Has that no part to play in safety? Should people not accept some responsibility for their own safety?

We live in a world in which people are less and less willing to accept responsibility for their actions. If a man commits a crime it is not his fault, but the fault of those who brought him up, or those who put him in a position in which he felt compelled to commit the crime. He should not be blamed, but offered sympathy. If someone is reluctant to work, he or she is no longer work-shy or lazy but a sufferer from chronic fatigue or some other recently discovered syndrome.

This attitude is parodied in the story of the modern Samaritan who found a man lying injured by the roadside and said, 'Whoever did this to you must be in need of help'. And in the story of the schoolboy in trouble who asked his father, 'What's to blame, my environment or my heredity?' Either way it was not him.

Many people react to this attitude by re-asserting that people do have free will and are responsible for their actions. A criminal may say that his crimes were the result of present or past deprivation, but most deprived people do not turn to crime.

One psychologist writes:

'... criminals aren't victims of upbringing; parents and siblings are victims of the individual's criminality ... Nor are criminals victims of their peer group ... young criminals-to-be choose their peers, not the other way around ... Drugs and alcohol ... are less the reason why some become criminals than the tool they use...to provide themselves the "courage" to hit a bank.'[1]

How do we reconcile these conflicting opinions?

We should distinguish between what we can expect from individuals and what we can expect from people *en masse*.

As individuals we should accept (and teach our children to accept) that we are responsible for our actions, otherwise we are mere computers, programmed by our genes, our parents, our environment or society at large. We must try to work safely, try not to forget, try to learn, do our best.

But as managers, dealing with large numbers of people, we should expect them to behave like average people — forgetting a few things, making a few mistakes, taking a few short cuts, following custom and practice, even indulging in a little petty crime when the temptation is great, not to a great extent but doing so to the extent that experience shows people have done in the past. Changing people, if it can be done at all, is a slow business compared with the time-scale of plant design and operation. So let us proceed on the assumption that people will behave much as they have done in the past.

Reichel, discussing crime in libraries, writes:

'Professors who assign research projects that require hundreds of students to use a single source in a library inevitably invite trouble in our competitive and permissive society, unless they make a co-ordinated effort to provide multiple copies of the source and to subsidize convenient photostat machine operating costs.'[15]

An old method for discouraging the theft of fire buckets is to use buckets with a rounded base. They are kept on a hook.

A number of three-pin plugs were stolen from plant instruments. The manager left a box of plugs in the control room with a notice saying 'Help yourself'. Some of them were taken but not all and the thefts stopped.

In the early 19th century, Bank of England notes were easily forged. Instead of making forgery difficult, the authorities tried to suppress it by savagery. More than 300 people were hanged, and many transported for the lesser crime of passing forged notes, before the notes were redesigned[17]. The cartoonist George Cruikshank protested against the penalties by drawing the cartoon note shown in Figure 13.1.

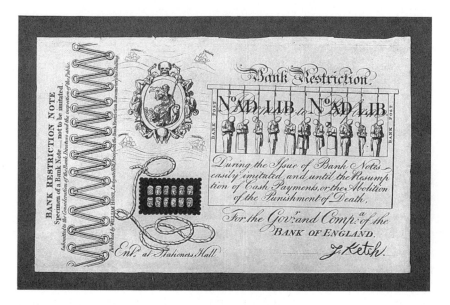

Figure 13.1 The cartoonist George Cruikshank protested against the severe penalties for forgery by drawing a Bank Restriction Note grimly decorated with skulls, gibbets and a hangman's noose. © The British Museum.

13.2 Legal views

UK law supports the view that we should expect people to behave in the future as they have behaved in the past, as these quotations from judges show:

'(A person) is not, of course, bound to anticipate folly in all its forms, but he is not entitled to put out of consideration the teachings of experience as to the form those follies commonly take.'[2] We could replace 'folly' by 'human error'.

'The Factories Act is there not merely to protect the careful, the vigilant and the conscientious workman, but, human nature being what it is, also the careless, the indolent, the weary and even perhaps in some cases the disobedient.'[3]

'The standard which the law requires is that (the employers) should take reasonable care for the safety of their workmen. In order to discharge that duty properly an employer must make allowance for the imperfections of the human nature. When he asks his men to work with dangerous substances he must provide appliances to safeguard them; he must set in force a proper system by which they use the appliances and take the necessary precautions, and he must do his best to see that they adhere to it. He must remember that men doing a routine task are often heedless of their own safety and may become slack about taking precautions.

'He must, therefore, by his foreman, do his best to keep them up to the mark and not tolerate any slackness. He cannot throw all the blame on them if he has not shown a good example himself.'[4]

'In Uddin v Associated Portland Cement Manufacturers Limited, a workman in a packing plant went, during working hours to the dust extracting plant — which he had no authority to do — to catch a pigeon flying around in the roof. He climbed a vertical steel ladder to a platform where he apparently leant over some machinery and caught his clothing on an unfenced horizontal revolving shaft, as a result of which he lost his arm. The trial judge found that the workman's action was the height of folly, but that the employer had failed to fence the machinery. The judge apportioned 20 per cent of the blame to the employer.

'In upholding the award, Lord Pearce, in his judgment in the Court of Appeal, spelt out the social justification for saddling an employer with liability whenever he fails to carry out his statutory obligations. The Factories Act, he said, would be quite unnecessary if all factory owners were to employ only those persons who were never stupid, careless, unreasonable or disobedient or never had moments of clumsiness, forgetfulness or aberration. Humanity was not made up of sweetly reasonable men, hence the necessity for legislation with the benevolent aim of enforcing precautions to prevent avoidable dangers in the interest of those subjected to risk (including those who do not help themselves by taking care not to be injured).

'Once the machinery is shown to be dangerous and require fencing, the employer is potentially liable to all who suffer from any failure to fence. And the duty is owed just as much to the crassly stupid as to the slightly negligent employee. It would not be in accord with this piece of social legislation that a certain degree of folly by an employed person should outlaw him from the law's protective umbrella.

'The accident in the pigeon case, it is true, would never have happened but for the unauthorised and stupid act of the employee. But then the accident would equally not have happened if the machinery had been properly fenced.'[5]

The judge added that the workman's actions were not 'actuated by benevolence towards the pigeon'. He wanted it for the pot.

The judge's comments suggest that many failures to work safely are deliberate. In fact I think more are due to a moment's forgetfulness. However, the law, like this book, is not primarily concerned with the reasons for failures to work safely but accepts that, for a variety of reasons, men will not always follow the rules and therefore designs and methods of operation should take this into account.

The quotations above were all made during claims for damages by injured people — that is, under the civil law. Under the criminal law, a manager can be prosecuted if he turns a blind eye to breach of the law — that is, if he sees someone working unsafely and says nothing.

'Where an offence under any of the relevant statutory provisions committed by a body corporate is proved to have been committed with the consent or connivance of, or to have been attributable to any neglect on the part of, any director, manager, secretary or other similar officer of the body corporate or a person who was purporting to act in any such capacity, he as well as the body corporate shall be guilty of that offence and shall be liable to be proceeded against and punished accordingly.'[6]

'"Connivance" connotes a specific mental state not amounting to actual consent to the commission of the offence in question, concomitant with a failure to take any step to prevent or discourage the commission of that offence.'[7] (See Section 3.6, page 66.)

In practice, in the UK, the company is usually prosecuted and prosecution of individuals is rare. It seems to occur only when there has been gross negligence, deliberate damage to safety equipment or failure to carry out duties which were clearly laid down as part of the job. For example, in 1996 an asbestos contractor who took no precautions to prevent the spread of asbestos was jailed for three months. This was the first time someone was imprisoned for a health and safety offence[23]. Later the same year, after a fatal accident, the managing director of a haulage company was imprisoned for a year for failing to provide adequate protective equipment and a safe system of work[24]. In 1998 a self-employed contractor was imprisoned for dumping asbestos; the company and its directors were fined[25].

As long as one does one's best and exercises reasonable care and skill, I do not think that there is any danger of prosecution. Lord Denning said in the Court of Appeal, 'The Courts must say firmly, that in a professional man, an error of judgement is not negligence.'[18] (Your employer might be prosecuted for not employing a more competent person, but that is another matter.)

If an engineer guesses the size of a relief valve instead of calculating it, or getting a competent person to calculate it, this is not an error of judgement but a failure to apply reasonable care and skill.

The Health and Safety at Work Act, Section 6(1) imposes duties on 'any person who designs, manufactures, imports or supplies any article for use at work'. Section 6(7) states that such liability 'shall extend only to things done in the course of a trade, business or other undertaking carried out by him (whether for profit or not) and to matters within his control'. This seems to exclude the employee designer.

According to Section 6 of the Health and Safety at Work Act 1974, the designer, manufacturer, supplier and user of equipment must ensure, so far as is reasonably practicable, that the equipment is safe and without risk to health when properly used. According to a 1987 amendment 'an absence of safety or a risk to health shall be disregarded in so far as it could not reasonably have been foreseen'[19].

It is interesting to note that the law on safety has a long history. The Bible tells us:

'When thou buildest a new house, then thou shalt make a battlement for thy roof, that thou bring not blood upon thine house, if any man fall from thence.'[8]

In the East the roofs of houses were (and still are) used as extra living space (Figure 13.2).

Figure 13.2 Roof of house in present-day Jerusalem showing 'battlements'

13.3 Blame in accident investigations

It follows from the arguments of this book that there is little place for blame in accident investigations, even when the accident is due to 'human error'. Everybody makes mistakes or slips and has lapses of attention from time to time and sometimes they result in accidents. Plants should be designed and operated so that these foreseeable errors do not result in accidents (or, if the consequences are not serious, the occasional accident can be accepted). People should not be blamed for behaving as most people would behave.

If we consider separately the various types of human error discussed in this book, there is no place for blaming the operator if the error is due to a slip or lapse of attention (Chapter 2), to lack of training or instructions (Chapter 3) or to lack of ability (Chapter 4). (Sometimes the employer can fairly be blamed for not providing a better system of work or better training or instructions or for not employing a more competent person.) Blame is relevant only when the person concerned has a choice and the deliberate decisions discussed in Chapter 5 are the only errors that could justify blame. Even here, the violation may have occurred because the rules were not clear, or the reasons for them had not been explained or someone turned a blind eye when they were broken in the past. Before blaming anyone we should answer the questions listed at the beginning of Chapter 5. Even if blame is justified it may not be wise, as it may result in people being less forthcoming in the future and make it harder for us to find out what happened (see Section 1.5, page 8).

Of course, if someone makes repeated errors, more than a normal person would make, or shows that he is incapable of understanding what he is required to do, or is unwilling to do it, then he may have to be moved.

Sometimes people are blamed because those in charge wish to deflect criticism from themselves on to a scapegoat (see Section 3.8, page 72). People may be blamed for a moment's slip or lapse of attention because a manager, perhaps unconsciously, wishes to divert attention from his own failure to provide a safe plant or system of work. In my experience, managers in the process industries, at least in the larger companies, do not often do this. After an accident they are usually willing to stand in a white sheet and admit that they might have done more to prevent it. The same is not true for all industries. In 1976 a mid-air collision in Yugoslavia killed 176 people. The air traffic controllers were overworked and the control equipment was poor. One controller went to look for his relief, who was late, leaving his assistant alone for eight minutes. The assistant made a slip which led to the accident. All the controllers on duty were arrested, eight were tried and the overworked assistant was sentenced to seven years' imprisonment. Following widespread protests he was released after two years[20].

If there is ever a martyrology of loss prevention, the Yugoslav controller should be the first candidate for inclusion. Another candidate for inclusion is the UK train driver who, in 1990, was sent to prison for passing a signal at danger[22] (see Section 2.9.3, page 38).

When companies try to get away from a blame culture, employees are sometimes so used to using blame as an explanation that they start to blame equipment.

13.4 Managerial wickedness

The reasons for human error that I have discussed in this book, and repeated in the last section, do not include indifference to injury as very few accidents result from a deliberate cold-blooded decision, by managers or workers, to ignore risks for the sake of extra profit or output. (I say 'cold-blooded' because in the heat of the moment we are all tempted to take a chance that, if we had time to reflect, we would recognize as unwise.)

For example, the report on the Clapham Junction railway accident[21] (see Section 6.5, page 118) says, '... I am satisfied that the errors which were made did not result from any deliberate decision to cut corners on safety,' and, 'the sincerity of the beliefs of those in BR at the time of the Clapham Junction accident ... cannot for a moment be doubted.'

The view that accidents are not due to managerial wickedness is not universally shared. Some writers seem to believe that it is the principal cause of accidents. They look on industrial safety as a conflict between 'baddies' (managers) and 'goodies' (workers, trade union officials, safety officers, and perhaps regulators). A book intended for trade union members says, 'Workers have seen how little their lives have often been valued.'[9]

This may have been true at one time, but not today. Managers are genuinely distressed when an accident occurs on their plant and would do anything to be able to put the clock back and prevent it. (A time machine such as Dr Who's Tardis would be a useful piece of equipment on every plant.) Accidents occur because managers lack knowledge, imagination and drive, do not see the hazards that are there and are subject to all the other weaknesses that beset human nature, but not because they would rather see people hurt than take steps to prevent them getting hurt.

It has been said that the three causes of accidents are:

- ignorance;
- apathy; and
- avarice.

In industry, I would agree about the first two but the third is the least important (see Section 6.6, page 120).

To conclude this section, here are two quotations from official reports. The first is from the Robens Report[10], which led to the 1974 Health and Safety at Work Act:

'The fact is — and we believe this to be widely recognized — the traditional concepts of the criminal law are not readily applicable to the majority of infringements which arise under this type of legislation. Relatively few offences are clear-cut, few arise from reckless indifference to the possibility of causing injury, few can be laid without qualification at the door of a particular individual. The typical infringement or combination of infringements arises rather through carelessness, oversight, lack of knowledge or means, inadequate supervision, or sheer inefficiency. In such circumstances the process of prosecution and punishment by the criminal courts is largely an irrelevancy. The real need is for a constructive means of ensuring that practical improvements are made and preventative measures adopted.'

The second is from the Aberfan Report[11]:

'... there are no villains in this harrowing story ... but the Aberfan disaster is a terrifying tale of bungling ineptitude by many men charged with tasks for which they were totally unfitted, of failure to heed clear warnings, and of total lack of direction from above. Not villains, but decent men, led astray by foolishness or ignorance or both in combination, are responsible for what happened at Aberfan.'

13.5 Managerial competence

If accidents are not due to managerial wickedness, *they can be prevented by better management*. The words in italics sum up this book. All my recommendations call for action by managers. While we would like individual workers to take more care, and to pay more attention to the rules, we should try to design our plants and methods of working so as to remove or reduce opportunities for error. And if individual workers do take more care it will be as a result of managerial initiatives — action to make them more aware of the hazards and more knowledgeable about ways to avoid them.

Exhortation to work safely is not an effective management action. Behavioural safety training, as mentioned at the end of the paragraph, can produce substantial reductions in those accidents which are due to people not wearing the correct protective clothing, using the wrong tools for the job, leaving junk for others to trip over, etc. However, a word of warning: experience shows that a low rate of such accidents and a low lost-time injury rate do not prove that the process safety is equally good. Serious process accidents have often occurred in companies that boasted about their low rates of lost-time and mechanical accidents (see Section 5.3, page 107).

Why then do published accident statistics say that so many accidents — over 50% and sometimes 80 or 90% — are due to 'human failing'?

There are several reasons:

(1) Accident reports are written by managers and it is easy to blame the other person.

(2) It is easier to tell a man to be careful than to modify the plant or method of working.

(3) Accidents are due to human failing. This is not untrue, merely unhelpful. To say accidents are due to human failing is like saying falls are due to gravity. It is true but it does not help us prevent them. We should list only those accident causes we can do something about (see Section 1.3, page 4).

(4) Sometimes, there is a desire to find scapegoats (see previous section).

So my counsel for managers is not one of comfort but one of challenge. You can prevent most accidents, not immediately, but in the long run, if you are prepared to make the effort.

Let me end this section with a few quotations. The first is from a union official:

'I place the blame for 99 per cent of all accidents fairly and squarely on the shoulders of management, directors, managers, foremen and chargehands, both in the past and right up to the present time.

'Unsafe to dangerous practices are carried out which anybody with an observant eye could see if they wished, and if they do see, do they do anything about it? Not until a serious accident happens and then the excuse is "It has never happened before. The job has always been done like this". The workman takes his cue from management. If the management doesn't care the workman doesn't care either until something happens.

'Sole responsibility for any accident should be placed fairly and squarely on the shoulders of the Departmental Manager. He should go about with his eyes open instead of sitting in his office, to be able to note unsafe and dangerous practices or work places, and get something done about them as soon as possible. If he gives instructions on these matters he should enforce them.'[12]

The second is from a safety engineer:

'... outstanding safety performances occur when the plant management does its job well. A low accident rate, like efficient production, is an implicit consequence of managerial control.

'Amplification of the desired managerial effect is more certain when managers apply the same vigorous and positive administrative persuasiveness that underlies success in any business function.'[13]

The third is a headline in an old copy of the *Farmers Weekly*:

TAIL BITING IN PIGS
FAULTS IN MANAGEMENT AND SUPERVISION

13.6 Possible and necessary

Figure 13.3 shows what can be achieved by determined management action. It shows how ICI's fatal accident rate (expressed as a five-year moving average) fell after 1968 when a series of serious accidents drew attention to the worsening performance[14].

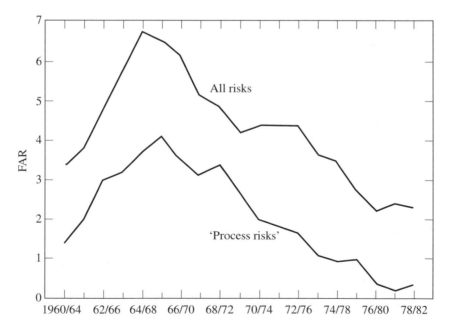

Figure 13.3 ICI's fatal accident rates (the number of fatal accidents in 10^8 working hours or in a group of 1000 men in a working lifetime) expressed as a five-year moving average, for the period 1960–1982 (from Reference 14).

231

I have heard colleagues suggest that the improved performance was due to Hazop, QRA or some other technique. In fact, we made many changes — we could not carry out controlled experiments — and we do not know which changes were the most effective. There was probably symbiosis between them.

Following the advice in this book calls for a lot of effort. Is it really necessary? A lot of the time it may not be. A lot of the time we can get by with a performance that is just adequate. Only in key situations are our best efforts really necessary. But we never know until afterwards which are the key situations. We never know when the others have failed and everything depends on us, when we are the last line of defence. The only safe way is to assume that our contribution is essential.

There are some things, such as turning lead into gold, we cannot do because there are no known methods. Preventing accidents is not one of them. The information is available, some of it in this book. What is often lacking is recognition of the need, knowledge of the actions needed and commitment. In this and other books I have tried to contribute towards the first two. Commitment is up to you.

References in Chapter 13

1. Page, J., 1984, *Science 84*, September 1984, 84.
2. Whincup, M., *The Guardian*, 7 February 1966. The quotation is from a House of Lords judgement.
3. A quotation from a judge's summing up, origin unknown.
4. A quotation from a judge's summing up which appeared in several newspapers in October 1968.
5. 'Justinian', *Financial Times*, June 1965.
6. Health and Safety at Work Act (1974), Section 37(1).
7. Fife, I. and Machin, E.A. (editors), 1982, *Redgrave's Health and Safety in Factories*, 2nd edition, page 16 (Butterworth, London, UK).
8. *Deuteronomy*, Chapter 22, Verse 8.
9. Eva, D. and Oswald, R., 1981, *Health and Safety at Work*, page 39 (Pan Books, London, UK).
10. *Safety and Health at Work — Report of the Committee 1970–1972* (the Robens Report), 1972, Paragraph 26.1 (HMSO, London, UK).
11. *Report of the Tribunal Appointed to Inquire into the Disaster at Aberfan on October 21st, 1966*, 1967, Paragraph 47 (HMSO, London, UK).
12. Hynes, F., August 1971, *Safety* (British Steel Corporation).
13. Grimaldi, J.V., 1966, *Management and Industrial Safety Achievement, Information Sheet No. 13* (International Occupational Safety and Health Information Centre (CIS), Geneva, Switzerland).

14. Hawksley, J.L., 1984, *Proceedings of the CHEMRAWN III World Conference (CHEMical Research Applied to World Needs), The Hague, 25–29 June 1984*, Paper 3.V.2.
15. Reichel, A.I., *Journal of Academic Librarianship*, November 1984, page 219.
16. Hidden, A. (Chairman), 1989, *Investigation into the Clapham Junction Railway Accident*, Paragraph 16.3 (HMSO, London, UK).
17. Hewitt, V.H. and Keyworth, J.M., 1987, *As Good as Gold* (British Museum, London, UK).
18. *The Times*, 6 December 1979.
19. Fife, I. and Machin, E.A. (editors), *Redgrave's Health and Safety in Factories* (Butterworths, London, UK, 1982, page 432 and Butterworth-Heinemann, Oxford, UK, 1987, page 339).
20. Stewart, S., 1986, *Air Disasters*, page 131 (Ian Allan, UK).
21. As Reference 16, Paragraphs 13.13 and 13.3.
22. *Daily Telegraph*, 4 September 1990.
23. *Health and Safety at Work*, 1996, 18(3): 5.
24. *Health and Safety at Work*, 1996, 18(12): 5.
25. *Health and Safety at Work*, 1998, 20(10): 4.

The adventures of Joe Soap and John Doe

14

'Today, a worker injury is recognised to be a management failure.'
P.L. Thibaut Brian, 1988, *Preventing Major Chemical and Related Process Accidents, Symposium Series No. 110*, page 65
(Institution of Chemical Engineers, Rugby, UK)

Joe Soap makes frequent mistakes — that is, he takes the wrong action as he has not been told what is the right action. In contrast, John Doe makes the little slips we all make from time to time. We could tell him to be more careful. Or we could make simple changes to plant design or methods of operation, to remove opportunities for error.

These cartoons originally appeared on page-a-month calendars. As a result no picture was displayed for so long that it became tatty or became part of the unnoticed background.

Figure 14.1

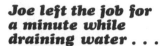

Joe left the job for a minute while draining water . . . . .

. and

Figure 14.2

Joe asked a fitter to do a job on tank F1374B

Figure 14.3

Joe saw a vent near a walkway...

Figure 14.4

Joe carried petrol .. **.. in a BUCKET !**

Figure 14.5

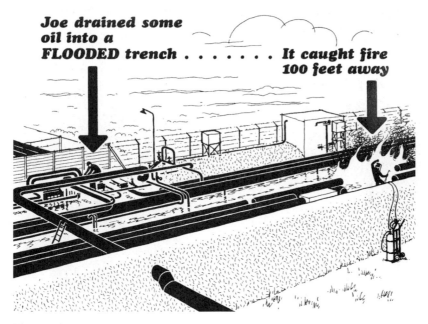

Figure 14.6

Figure 14.7

Joe tested the inside of a vessel with a gas detector and—getting a zero reading—allowed welding to go ahead

Oils with a flash point above atmospheric temperature are not detected by gas detectors

HEAVY OIL

Figure 14.8

Joe left off his goggles and stood too close to take a sample

Figure 14.9

Joe added hot oil (over 100°C) to a tank containing some water

Figure 14.10

Joe tied a plastic bag over the vent to keep the tank clean

Figure 14.11

Figure 14.12

Figure 14.13

Further Adventures of Joe Soap

Once again poor Joe makes a mistake every month.

All these incidents happened to someone — though not to the same man — and may happen to us unless .....

Figure 14.14

Figure 14.15

Joe ignored an instrument reading which 'could not possibly' be correct...

It WAS

Figure 14.16

Joe tested the line at 10 a.m.

The welder started at 3 p.m.

Test JUST BEFORE welding

Figure 14.17

Joe let the tractor leave and then emptied the rear compartment first

Figure 14.18

Joe disconnected a flex before releasing the pressure

Figure 14.19

Joe let a furnace tube get 150°C too hot for 8 hours

The tube burst after half its normal life

Figure 14.20

Joe left a portable light face down on a staging

Figure 14.21

Joe put his oily overalls on a warm line to dry

Figure 14.22

Joe put some cylinders into a closed van .....!

Figure 14.23

Figure 14.24

Figure 14.25

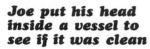

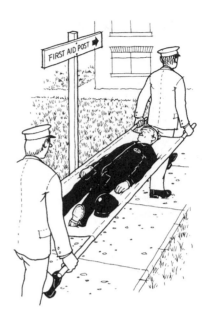

Figure 14.26

Now learn from the adventures of John Doe …

Figure 14.27

Figure 14.28

Figure 14.29

Figure 14.30

Figure 14.31

249

Figure 14.32

Figure 14.33

Figure 14.34

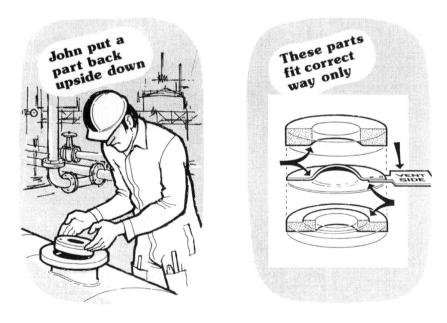

Figure 14.35

Figure 14.36

Figure 14.37

Figure 14.38

Figure 14.39

Some final thoughts

15

'The best that can be hoped for from the ending is that sooner or later it will arrive.'
N.F. Simpson, *A Resounding Tinkle*

'Never attribute to malice or other deliberate decision what can be explained by human frailty, imperfection or ignorance.'
After H.S. Kushner[1]

I have tried to show that human errors are events of different types (slips and lapses of attention, mistakes (including ignorance of responsibilities), violations, errors of judgement and mismatches, made by different people (managers, designers, operators, construction workers, maintenance workers and so on) and that different actions are required to prevent them happening again — in some cases better training or instructions, in other cases better enforcement of the rules, in most cases a change in the work situation (see Table 1.1, page 10). Almost any accident can be said to be due to an error by someone and the use of the phrase discourages constructive thinking about the action needed to prevent it happening again; it is too easy to tell someone to be more careful. Human error is one of those phrases, like 'Act of God' and 'cruel fate' that discourage critical thought. These terms imply that we can do little or nothing about natural disasters. Human error implies that we can do little more than tell people to take more care.

Wrigglesworth wrote (in 1972)[2]:

'Accidental death is now the only source of morbidity for which explanations that are essentially non-rational are still socially acceptable. Suggestions that 'it was an Act of God' or 'it was a pure accident' or 'it was just bad luck' are still advanced as comprehensive explanations of the causes of accidents and are accepted as such even by professionally trained persons who, in their own fields of activity, are accustomed to apply precise analytical thinking. This approach is typical of the fables, fictions and fairy tales that mark our community approach to the accident problem.'

There has been much progress since this was written but human error is still quoted, more often than Acts of God or bad luck, as the cause of accidents as if it was a sufficient explanation. If human error covers different actions, by different people, requires such different remedies and discourages thought about these remedies, is the concept a useful one?

In a book on radio[3] that I was given as a school prize in 1940 there is a chapter on the ether. It says, 'The whole of space is filled with ether — a continuous very dense and very elastic medium.' The author surmises that, 'the ether itself consists of very fine particles which, though practically incompressible, can easily be shifted in respect to each other.'

Today no-one would write like this. We have realized that the concept of the ether is unnecessary. It does not explain anything that cannot be explained without it.

Similarly, biologists have abandoned the idea of protoplasm as the physical basis of life, something which permeates inanimate structures and gives them vitality. To quote the Medawars, 'Today the word "protoplasm" is falling into disuse, a fate that would certainly not have overtaken it if it had served any useful purpose, even as a first approximation to reality.'[4]

Perhaps the time has come when the concept of human error ought to go the way of phlogiston, the ether and protoplasm. Perhaps we should let the term 'human error' fade from our vocabularies, stop asking if it is the 'cause' of an accident, and instead ask what action is required to prevent it happening again. Perhaps 'cause' as well as 'human error' ought to go the way of phlogiston, ether and protoplasm. [Ether, incidentally, has twice been discarded as a concept of no value. It originally meant a fifth element, in addition to earth, air, fire and water, something halfway between mind and matter that could convert thoughts into physical events[5].]

According to Dewey[6]:

'... intellectual progress usually occurs through shear abandonment of questions together with both of the alternatives they assume — an abandonment that results from their decreasing vitality ... We do not solve them: we get over them. Old questions are solved by disappearing, evaporating, while new questions corresponding to the changed attitude of endeavour and preference take their place.'

References in Chapter 15

1. Based on a quotation from Kushner, H.S., 1996, *How Good Do You Have to Be?*, page 109 (Brown, Boston, MA, USA).
2. Wrigglesworth, E.C., *Occupational Safety and Health*, April 1972, page 10.
3. Stranger, R., 1939, *The Outline of Wireless*, page 202 (Newnes, London, UK).
4. Medawar, P. and J., 1983, *Aristotle to Zoos*, page 220 (Oxford University Press, Oxford, UK).

5. Miller, J., 1978, *The Body in Question*, page 290 (Cape, London, UK).
6. Dewey, J., 1910, The influence of Darwinism on philosophy, quoted in Gardner, M. (editor), 1985, *The Sacred Beetle and Other Great Essays in Science*, 2nd edition, page 20 (Oxford University Press, Oxford, UK).

Postscript

'... there is no greater delusion than to suppose that the spirit will work miracles merely because a number of people who fancy themselves spiritual keep on saying it will work them.'
L.P. Jacks, 1931, *The Education of the Whole Man*, 77 (University of London Press) (also published by Cedric Chivers, 1966)

Religious and political leaders often ask for a change of heart. Perhaps, like engineers, they should accept people as they find them and try to devise laws, institutions, codes of conduct and so on that will produce a better world without asking for people to change. Perhaps, instead of asking for a change in attitude, they should just help people with their problems. For example, after describing the technological and economic changes needed to provide sufficient food for the foreseeable increase in the world's population, Goklany writes[1]:

'... the above measures, while no panacea, are more likely to be successful than fervent and well-meaning calls, often unaccompanied by any practical programme, to reduce populations, change diets or life-styles, or embrace asceticism. Heroes and saints may be able to transcend human nature, but few ordinary mortals can.'

Reference
1. Goklany, I.M., 1999, in Morris, J. and Bate, R. (editors), *Fearing Food: Risk, Health and Environment*, page 256 (Butterworth-Heinemann, Oxford, UK).

Appendix 1 – Influences on morale

In Section 4.3 (page 85) I suggest that some people may deliberately but unconsciously injure themselves in order to withdraw from a work situation which they find intolerable[2] (or, having accidentally injured themselves, use this as an excuse for withdrawing from work). If a group of employees feel oppressed they may strike. If an individual employee feels oppressed he may drag his feet, not comply, feign ignorance, swing the lead, make the most of any injury and perhaps even injure himself. According to Husband and Hinton, '[some] children react to a difficult family situation by hurting themselves'[2]. (See Reference 10 of Chapter 4.) Perhaps they do not get on with their fellow workers; perhaps their job does not provide opportunities for growth, achievement, responsibility and recognition. The following notes develop this latter theme a little further, but I want to emphasize that, as stated in Section 4.3, it is not a major contribution to accident rates. It may contribute more to high sickness absence rates.

I want to avoid technical terms as far as possible but there are two that must be used:

Hygienic (or maintenance) *factors and motivators*

Hygienic factors include rates of pay, fringe benefits, working conditions, status symbols and social grouping. *They have to reach a minimum acceptable standard or employees will be dissatisfied and will not work well, but improving these factors beyond the minimum will not make anyone work better.* The minimum acceptable standard varies from time to time, from place to place and from person to person.

For example, people will not work as well as they might if the pay is low, there is no security, the workplace, canteens and washrooms are filthy and they are treated like dirt. All these factors must be brought up to an acceptable standard before we can hope to get anyone to work well. But improving them alone will not persuade people to work; some companies have given all

— high pay, security, pensions, good holidays, fine canteens, sports fields, clubs, children's outings — but still staff and payroll remain 'browned-off and bloody-minded'.

What's gone wrong?

The answer, according to Herzberg[1], is that to get people to work well we must:

- First, bring the hygienic factors up to scratch (if they are below it).
- Second, *fulfil people's need for growth, achievement, responsibility and recognition. Give them a job with a definite aim or objective* rather than a collection of isolated tasks, involve them in deciding what that objective should be and how it is achieved, give them as much freedom as possible to decide how they achieve that objective, show them how their work fits into the wider picture, tell them when they have done a good job and make it clear that there are opportunities for promotion.

This theory explains why we are so well motivated in wartime (the aim or objective is obvious) and why process workers, who have the feeling they are driving the bus, are usually better motivated than maintenance workers. The theory is easier to apply to staff jobs than payroll ones but nevertheless let us try to apply it to the floor-sweeper. The object is obvious and unexciting. What about involvement? It is usual to give the floor-sweeper a schedule, drawn up by someone else, which says which areas are to be swept each day. Why not involve the floor-sweeper in drawing up the schedule? He knows better than anyone which areas need sweeping daily, which weekly. How often does the manager (or anyone else on the plant) thank him for keeping it so clean and make him feel he is 'in charge of the plant'. (He forgets to add 'for cleaning' when he repeats it to his wife.)

Some men attach more importance to hygienic factors than others. The 'maintenance seeker' is seeking all the time for:

- more salary;
- better working conditions;
- more 'perks';
- more status trappings;
- more security;
- less supervision.

He knows all the rules and all the injustices, little and big. If asked about his job, he describes the working conditions, perks, etc.

The 'motivation seeker' on the other hand is motivated by the work itself rather than the surroundings. He is seeking all the time to complete more jobs, solve more problems, take more responsibility, earn recognition.

But people are not born maintenance seekers or motivation seekers. Surroundings rich in opportunities for satisfying motivation needs breed motivation seekers and vice versa.

At one time I used to attend regular meetings of works managers at which the lost-time accident rates were displayed. If a works had a high accident rate, the works manager would often explain that there had not really been an increase in accidents but that a number of men had decided to take time off after suffering minor injuries. The production director would then ask, 'Why do your men lose time over minor injuries when the men in other works do not?'

References in Appendix 1

1. Herzberg, F., 1968, *Work and the Nature of Man* (Staples Press, London, UK).
2. Husband, P. and Hinton, P., *Care in the Home*, January 1973, page 13.

Appendix 2 – Some myths of human error

'Often, the ideas put forward made (and make) perfect sense.
Their only problem is that they are wrong.'
Steve Jones[1]

In my book *Dispelling Chemical Engineering Myths*[2] I list nearly a hundred beliefs about technology, management and the environment which are not wholly true, though there is often some truth in them. The following are some myths on human error. Most of the points made have been discussed already in this book but it may be useful to have them collected together. The first six are discussed in greater detail in Reference 2.

1. Most accidents are due to human error

This is true, if we include errors by designers and managers. But this is not what people usually mean when they say that most accidents are due to human error. They usually mean that most accidents are caused by people at the sharp end: the operator who opens the wrong valve, the railway engine driver who passes a signal at danger, the airline pilot who pulls the wrong lever. In fact, they are at the end of a long line of people all of whom could have prevented the accident.

Figure A2.1 (page 262) summarizes an example which combines bits from several incidents. A bellows was incorrectly installed so that it was distorted. After some months it leaked and a passing vehicle ignited the escaping vapour. Damage was extensive as the surrounding equipment had not been fire-protected, to save cost. The leak would not have occurred, or the damage would have been less, if:

- Bellows were not allowed on lines carrying hazardous materials.
- The use of bellows had been questioned during design. A hazard and operability study would have provided an opportunity to do so.
- The fitter who installed the bellows had done a better job. Did he know the correct way to install a bellows and the consequences of incorrect installation?
- There had been a thorough inspection after construction and regular inspections of items of equipment whose failure could have serious consequences.
- Everyone had kept their eyes open as they walked round the plant.
- Gas detectors and emergency isolation valves had been installed.

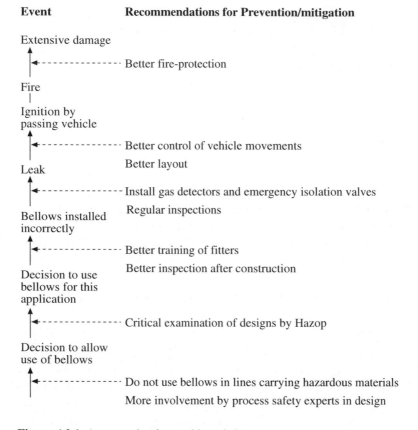

Event **Recommendations for Prevention/mitigation**

Extensive damage
↑ ⟵-------------- Better fire-protection

Fire

Ignition by
passing vehicle
↑ ⟵-------------- Better control of vehicle movements
Leak Better layout
↑ ⟵-------------- Install gas detectors and emergency isolation valves
Bellows installed Regular inspections
incorrectly
↑ ⟵-------------- Better training of fitters
Decision to use Better inspection after construction
bellows for this
application
↑ ⟵-------------- Critical examination of designs by Hazop
Decision to allow
use of bellows
↑ ⟵-------------- Do not use bellows in lines carrying hazardous materials
  More involvement by process safety experts in design

Figure A2.1 An example of an accident chain

- The plant had been laid out so that vehicles delivering supplies did not have to pass close to operating equipment. Did the designers know that diesel engines can ignite leaks of flammable vapour?
- There had been better control of vehicle movements.
- The fire protection had been better.
- An expert in process safety was involved during design, as he would have drawn attention to many of these items.

There were thus at least ten ways in which the chain of events leading to the damage could have been broken and many people who could have broken it. They were all responsible to some extent and it would be wrong and unfair to pick on one of them, such as the fitter who installed the bellows incorrectly, and make him the culprit. (See the extract from the Robens Report in

262

Section 13.4, page 228.) Unfortunately, after a serious accident the press think there must be a 'guilty man' who is responsible. There are also managers who lay a smoke screen over their own responsibility by blaming workmen who broke a rule or did a bad job, though they never enforced the rule or checked the workmanship.

2. Accidents are caused by people so we should eliminate the human element whenever possible

We cannot eliminate the human element. If we automate an operation we are no longer dependent on the operator but we are now dependent on the people who design, manufacture, install, test and maintain the automatic equipment. They also make errors. It may be right to make the change as these people have more time and opportunities than operators to check their work, but we must not kid ourselves that we have removed the human element (see Section 7.5.2, page 143).

3. Automatic equipment is unreliable; the more we install, the more failures we get, so on the whole, it is better to rely on people

This is the opposite of the second myth. Up to a point, the reliability of a trip can be increased to any desired level by more frequent testing or by redundancy or diversity (that is, by duplicating components or providing other components capable of carrying out the same function). Voting systems can prevent spurious trips. However, automatic equipment is not always more reliable than people. Each application should be considered on its merits (see Section 7.5.2, page 143).

4. If operators have adequate time — say, 30 minutes — to carry out a simple task, such as closing a valve after an alarm sounds, then we can assume that they always will always do so and there is no need for automatic equipment?

This is true only when:

- The alarm can be heard.
- The operator knows what he is expected to do.
- The equipment is clearly labelled.
- The operator knows why he is expected to do it. If not, he may consider it unimportant.
- The task is within the operator's physical and mental ability (for example, the valve is not too stiff or out of reach).
- A blind eye has not been turned to past non-compliance.
- The operator does not have too many more urgent tasks to carry out first.

5. The best workers are those who get things done quickly

Not always; when hardware, methods of operation or organizations have to be changed, the 'go-getter' who rushes may repent at leisure. The best workers are those who first look critically and systematically at proposed changes to see if there will be any undesirable consequences. There often are[3,4].

6. The technical people know their jobs. The safety adviser in a high-technology industry can leave the technology to them and attend to human problems.

This myth is believed by those company heads who make the safety advisers responsible to the human resources manager (and by those safety advisers who do not understand the technology and take an interest in simple mechanical accidents only). Unfortunately, experience shows that many technical people are not familiar with the accidents of the past and the actions necessary to prevent them happening again. It should be one of the safety adviser's jobs to see that they are trained when they arrive and continually reminded and updated throughout their careers.

7. Following detailed rules and regulations can prevent accidents

In some countries, including the US, this is widely believed and the authorities write books of detailed regulations that look like telephone directories though they are less interesting to read (and probably consulted less often). In the UK, since 1974, there has been a general obligation to provide a safe plant and system of work and adequate instruction, training and supervision, so far as is reasonably practicable. The regulations set goals that must be achieved but do not specify exactly how this must be done, though advice and codes of practice are available.

The UK system is better because:

- It is impracticable to write detailed rules for complex and changing technologies.
- Codes and advice can be updated more quickly than regulations.
- Employers cannot say, 'It must be safe as I am following all the rules.'
- Regulators do not have to prove that a rule has been broken. It is sufficient for them to show that an employer has not provided a safe plant or system of work.

8. The English language (unlike French) has few irregular verbs

It does have a number that are not recognized as such. For example:

- I am firm.
- You are stubborn.
- He is pigheaded.

So far as safety is concerned, we have:

- I show initiative.
- You break the rules.
- He is trying to wreck the job.

9. We have done all we can to improve equipment. From now on we should concentrate on changing human behaviour and on management.

This is often heard but is not true. We are still producing poor designs because the designers are not aware of the alternatives. Sometimes they are not aware of cheap minor changes that could improve safety at little or no cost. For example, using different hose connectors on compressed air and nitrogen lines (see Section 2.6, page 31) or using valves whose position (open or closed) is visible at a glance. Sometimes they are not aware of fundamentally different design options such as inherently safer designs (see Chapter 8, especially Section 8.7, page 162).

10. If we reduce risks by better design, people compensate by working less safely. They keep the risk level constant.

There is some truth in this. If roads and cars are made safer, or seat belts are made compulsory, some people compensate by driving faster or taking other risks. But not all people do, as shown by the fact that UK accidents have fallen year by year though the number of cars on the road has increased. In industry many accidents are not under the control of operators at all. They occur as the result of bad design or ignorance of hazards.

11. Accidents are due to either human failure or equipment failure — that is, unsafe acts or unsafe equipment

This used to be a common belief among old-style safety officers. In fact, most accidents could have been prevented in many ways and many people had the opportunity to prevent them, as the case histories in this book have shown. An accident report may tell us more about the culture of a company and the beliefs of the writer than about the most effective means of prevention. In some companies the report points the finger at the person at the end of the chain, who opened the wrong valve. He requires extra training or reprimanding. They might suggest improving the labelling (though there may be no system for checking that labels have not vanished). In other companies they ask why there was no trip, interlock or relief valve to prevent such a simple slip having such serious consequences. In only a few companies do people ask if the hazard could be removed and why that was not done.

Consider a simple example: someone falls into a hole in the road. At one time the man or men who dug the hole would simply have been blamed for not fencing it. Today many companies are less simplistic. They ask if responsibilities were clear, if the fencing was readily available and everyone knew where it was kept. They also ask if holes had been left unprotected in the past and if no-one had said anything. In only a few companies, however, will someone ask why it was necessary to dig a hole in the road. Why not run pipes and cables above ground where access is better and corrosion less? If it was essential to run a new cable or pipe under the road, why not bore a hole under the road? Why not install conduits under the roads when plants are built? We can't fall into holes that aren't there.

12. The result of an accident measures the degree of negligence

This is widely believed by the general public and the police, though it is obviously untrue. If someone throws a scaffold pole off the top of a structure, instead of carrying it downstairs, this might kill someone, injure someone, cause damage or have no effect, but the degree of negligence is the same. The reaction of other people, including the police, would however be very different.

13. In complex systems, accidents are normal

In his book *Normal Accidents*[5], Perrow argues that accidents in complex systems are so likely that they must be considered normal (as in the expression SNAFU — System Normal, All Fowled Up). Complex systems, he says, are accident-prone, especially when they are tightly-coupled — that is, changes in one part produce results elsewhere. Errors or neglect in design, construction, operation or maintenance, component failure or unforeseen interactions are inevitable and will have serious results.

His answer is to scrap those complex systems we can do without, particularly nuclear power plants, which are very complex and very tightly-coupled, and try to improve the rest. His diagnosis is correct but not his remedy. He does not consider the alternative, the replacement of present designs by inherently safer and more user-friendly designs (see Section 8.7 on page 162 and Reference 6), that can withstand equipment failure and human error without serious effects on safety (though they are mentioned in passing and called 'forgiving'). He was writing in the early 1980s so his ignorance of these designs is excusable, but the same argument is still heard today.

14. To protect people from a deluge of time-wasting, unnecessary paper we should tell them only what they need to know

In some organizations employees are given the information they need to do their job, but no more. This system has snags. We don't know what we need to know until we know what there is to know.

New ideas come from a creative leap that connects disjointed ideas, from the ability to construct fruitful analogies between fields[7]. If our knowledge is restricted to one field of technology or company organization we will have few new ideas. Engineers need to build up a ragbag of bits and pieces of knowledge that may one day be useful. They will never do this if they are restricted (or restrict themselves) to what they need to know.

Creative people often 'belong to the wrong trade union'. If a civil engineer, say, has a new idea in, say, control engineering, the control engineers may be reluctant to accept it.

15. If the answer to a problem is correct, the method used to obtain it will be correct

The following is said to have been set as an examination question in the days when there were 20 shillings (s) in the pound and 12 pence (d) in the shilling[8].

Make out a bill for the following:

¼ lb butter	@ 2s 10d per lb
2 ½ lb lard	@ 10d per lb
3 lb sugar	@ 3 ¼d per lb
6 boxes matches	@ 7d per dozen
4 packets soap flakes	@ 2½d per packet

One boy added up the prices on the right and got the correct answer (4s 8 ¾d).

A more technical example: if we assume that the Earth absorbs all the radiation from the sun that falls on it then the average surface temperature works out as about 7°C, in good agreement with experience. However, about a third of the radiation is reflected by the clouds and if we allow for this the average temperature becomes −18°C. The heat produced by radioactivity makes the actual temperature higher[9].

16. The purpose of an accident investigation is to find out who should be blamed

This is widely believed by the public at large, newspaper reporters and the police. If you have got this far and still believe it, I have failed in my task.

References in Appendix 2

1. Jones, S., 1996, *In the Blood*, page 12 (Harper Collins, London, UK).
2. Kletz, T.A., 1966, *Dispelling Chemical Engineering Myths* (Taylor and Francis, Washington, DC, USA).
3. Sanders, R.E., 1999, *Chemical Process Safety: Learning from Case Histories* (Butterworth-Heinemann, Boston, MA, USA).
4. Kletz, T.A., 1998, *What Went Wrong? — Case Histories of Process Plant Disasters*, 4th edition, Chapter 2 (Gulf Publishing Company, Houston, Texas, USA).
5. Perrow, C., 1984, *Normal Accidents* (Basic Books, New York, USA).
6. Kletz, T.A., 1998, *Process Plants: A Handbook for Inherently Safer Design*, 2nd edition (Taylor and Francis, Philadelphia, PA, USA).
7. Gould, S.J., 1983, *The Panda's Thumb*, page 57 (Penguin Books, London, UK).
8. Maxwell, E.A., 1959, *Fallacies in Mathematics*, page 9 (Cambridge University Press, Cambridge, UK).
9. de Grasse Tyson, N., *Natural History*, April 1999, page 92.

Appendix 3 – Some thoughts on sonata form

Listening to music one evening, after working on this book during the day, it struck me that its argument could be summarized in sonata form.

For those unfamiliar with musical terms, the structure of sonata form, widely used in symphonies, concerti and quartets as well as sonatas, is as follows:

- A theme A is stated.
- It is followed by a contrasting theme B.
- Next, there is a development section in which both themes are developed but one may predominate. This section can be as short as a few bars or can take up most of the movement.
- Finally the themes are repeated, in a different key.
- Bridge passages may connect the different sections.

The argument of this book may be expressed as follows:

Theme A
We can prevent human errors by getting people to behave differently.

Contrasting theme B
It is more effective to change the task than change the person.

Bridge passage
We need to consider the different sorts of human error separately.

Development 1st part
Some errors (mistakes) are due to poor training or instructions: the person who made the error did not know what to do or how to do it. At first theme A is dominant: we can get people to behave differently by providing better training or instructions. But then theme B appears: perhaps we can make the task simpler and then we won't need so much training or such detailed instructions.

Development 2nd part
Some errors (violations or non-compliance) are the result of someone deciding not to follow instructions or recognized good practice. Again theme A is at first dominant: we can persuade people to follow instructions by explaining the reasons for them and, if that does not work, imposing penalties. But then theme B appears again: perhaps we should make the task easier, perhaps we ignored previous violations and thus implied that the tasks (or the way they are done) are not important.

Development 3rd part
Some errors (mismatches) are beyond the mental or physical ability of the person asked to do it, perhaps beyond anyone's ability. Theme B is dominant: we should make the task easier. There is just a suggestion of theme A: in some cases we may have to ask a different person to carry out the task in future.

Development 4th part
Many errors are the result of slips and lapses of attention. Theme A starts faintly, someone telling us to take more care, pay more attention. Then theme B interrupts, loudly and clearly telling us that we all make slips from time to time, particularly if we are busy, tired, distracted or under stress and there is little that we can do to prevent them. We should redesign tasks so that there are fewer opportunities for error (or so that the results of errors are less serious or recovery is possible).

Recapitulation
The two themes are repeated but now theme B is the main one while theme A, though still present, is now less assertive.

Further reading

Chapanis, A., 1965, *Man-machine Engineering* (Tavistock Press, London, UK).
The theme is very similar to the theme of this book but the examples are taken from mechanical and control engineering rather than the process industries.

Reason, J. and Mycielska, K., 1982, *Absent-minded? The Psychology of Mental Lapses and Everyday Errors* (Prentice-Hall, Englewood Cliffs, New Jersey, USA) (263 pages).
An account of the mechanisms underlying everyday and more serious slips. For a shorter account see Reason, J., 1984, Little slips and big disasters, *Interdisciplinary Science Reviews*, 9(2): 179–189.

Reason, J., 1990, *Human Error* (Cambridge University Press, Cambridge, UK) (302 pages).

Edwards, E. and Lees, F.P., 1973, *Man and Computer in Process Control* (Institution of Chemical Engineers, Rugby, UK) (303 pages).

Ergonomic Problems in Process Operation, Symposium Series No. 90, 1984 (Institution of Chemical Engineers, Rugby, UK) (229 pages).
A collection of papers on the role of the operator, interface design and job design.

Health and Safety Executive, 1989, *Human Factors in Industrial Safety* (HMSO, London, UK) (17 pages).
A brief account of the actions needed.

Brazendale, J. (editor), 1990, *Human Error in Risk Assessment, Report No. SRD/HSE/R510* (UK Atomic Energy Authority, Warrington, UK) (68 pages).
A brief report on methods of classification and prediction.

Center for Chemical Process Safety, 1994, *Guidelines for Preventing Human Error in Chemical Process Safety* (American Institute of Chemical Engineers, New York, USA) (390 pages).

A Manager's Guide to Reducing Human Errors: Improving Human Performance in the Chemical Industry, 1990 (Chemical Manufacturers Association, Washington, DC, USA) (63 pages).

Petroski, H., 1982, *To Engineer is Human — The Role of Failure in Successful Design* (St Martins Press, New York, USA) (245 pages).

Petroski, H., 1994, *Design Paradigms — Case Histories of Error and Judgement in Engineering* (Cambridge University Press, Cambridge, UK) (209 pages).

Improving Maintenance — A Guide to Reducing Error, 2000 (HSE Books, Sudbury, UK) (74 pages).

Index

A

Aberfan   184, 229
ability, lack of   78–94
Aborigines   123
access   79
accident investigation   122–123, 227
accident statistics   1, 87
accident–prone   12, 85–88
acetone   199
acids   45, 53, 58, 192–194, 199
'Acts of God'   254
agitation   32, 53
aircraft   9, 17, 41, 146–147, 161, 213–214, 227
Aisgill   38
alarm overload   80–82
alarms   21–22, 53, 58, 65–66, 74, 80–81, 100–101, 105, 125, 130, 149–150, 155, 204–205, 207, 210–211, 263
alertness   83–84, 198, 215
alienation   109–110
amateurism   115
ambulances   83
Amish   110
ammonia   169, 190
anaesthetics   31, 35
appearance, incorrect   34
asbestos   225
assembly   121
    incorrect   157, 161
    probability of errors   146
attitudes   75, 257

audits   2, 99, 118, 125, 129, 155, 178, 189
Australia   123, 148
auto–pilot   11–12
automatic train protection   39–40
automation   83–84, 104, 116, 130–131, 191, 253, 263
    and people compared   84, 143, 207–208, 211

B

back injuries   75
batch processes
    errors, probability of   140, 143–144
behavioural safety training   12, 75, 176, 229
behavioural science techniques   107–109
bellows   150, 159, 169, 173, 261–262
Belvoir Castle   110
beverage machines   25–26, 144–146, 150
Bhopal   121
Bible   74, 226
bicycles   85
'black book'   53
black box   207, 209
blame   8–9, 37, 97–98, 104, 113, 182, 213, 221, 225–229, 263, 266–267
blind eye   65, 99, 104, 119, 124, 225, 263
boilers (see also waste heat boilers)   17, 84, 105, 155, 179, 184–185
bolts   175
botching   63, 169

box girder bridges 57
brakes 200
breathing apparatus 62, 78, 161, 179, 182
bridges 57
bursting discs 251
buttons 25–29, 144–146, 250
  probability of pressing errors 144–146
Byng, Admiral 74

C

cable railways 9
cables 176
calculation errors 32–34, 213
cars 6, 16, 43, 63–64, 105, 151, 265
causes
  underlying 114–120, 122–123
  versus actions 255
caustic soda 193
chance 86–87
Channel Tunnel 65–66
Chazops 209
check–lists 16
checks 99, 104–107, 117–118, 147–149, 155, 169, 181, 196, 201
  weakness of 119
chemical burns 20, 58, 182–183
Chernobyl 52, 101–102
Chinley 38
chlorine 62
chloroform 31
chokes 50, 53–55, 91
clamps 43
Clapham Junction 45, 118–119, 228
classification systems 4, 7, 26
cleaning 193
clocks 43–44
clumsiness 79
cognitive dissonance 94
cokers 89–90
colliery tips 184

colour–coding 22–24, 26, 30, 32, 74, 175, 179, 212
column supports 91–92
common law 76
common mode failure 22
communication 119, 124, 208–210
  by radio 199
  verbal 198–199
  written 194–197
complexity 22, 121, 163, 266
compressed air 61, 161, 179, 194, 200, 209, 265
compressed gases 72
  power of 54
compressor houses 170
compressors 78, 169–170
computer displays 211
computers 8, 81, 203–218
  calculations by 217
  capability exaggerated 207, 215
  errors in entries 212–214, 216
  hardware failures 204–205
  responses to 210–213
  retention of data 211
  software errors 205–206
  specification errors 208–210
condensate 198
connections, wrong 31
connectors for hoses 31–32, 187–189, 193–194
conspiracy 94
construction 57, 81–82, 157, 168–173, 261
contamination 192
contractors 56, 64, 110, 115, 165, 172
contradictory instructions 65–66, 73
control rooms 27, 131–132
corrosion 104, 157, 201
  of procedures 4
costs 40, 113, 121, 125, 185
cranes 17, 27, 57, 78, 84
creativity 267
crime 221–226

culture   88, 109, 112, 176,
                          258–260, 265
cylinders   62, 78, 177–179, 245

D

databases   124–125
dates   214
dead–ends   169
debottlenecking   164
defence in depth   79
demanning (see manpower,
  reductions in)
demolition   44
design   2–4, 12, 16, 20–23, 32–34, 41,
          43, 57, 66, 80–81, 91, 115,
          117–118, 154–166, 170, 173, 265
  alternatives   23–24, 121, 162–165
  'build and maintain'   185
design liaison   164–165, 172
diagnosis   48, 52, 94
diesel engines   90, 115
dimensions, estimation of   85
directors, responsibilities of   112–125
discretion   73
discussions   50, 56, 71, 84, 98, 103, 155
distillation columns   17, 54, 132
distractions   26, 130, 146
documents, ancient   93
dolls   3
domestic accidents   43–44, 166, 240
drain valves   17–21, 63, 183, 235, 253
drowning   166
drums   30, 56, 64
duplication, effect on
  probability of error   147–149
dusts   61, 100, 201

E

education   48–76, 124
  of managers   113, 125
Einstellung   88
electric plugs   74

electrical area classification   164
electrical equipment   64, 79, 148, 164,
                          178, 204, 244
electricity, isolation of   17, 148
emergency evacuation   85
entry to confined spaces   71, 179, 196,
                          247, 251
equipment
  failures   149–150
  flameproof   178
  identification of   180, 187–189, 235,
                          248, 250
  opening of   13–16, 209
errors (see also train driver errors)
  in batch processes   140, 143–144
  by managers   112–125
  definition of   v
  medical   31, 34–35, 214
  in numbers   30–31
  measurement   146
  opportunities for   3
  probability of   6, 16, 26, 90,
                          120, 129–152
    examples   131–146
    limitations   129–130, 147–151
    reasons for   130–131
  in software   205–206
  recovery from   27
escalators   116
ether   255
ethylene   31–32, 91, 114, 163
ethylene oxide   163
events, rare, detection of   82
everyday errors   12
everyday life   11
evidence ignored (see mind–sets)
expectations   84–85
experiments   102
experts and amateurs   115
explosion venting   170
explosions   31–32, 58, 61, 64, 66–67, 71,
          80, 90, 100, 105, 109, 114, 120,
          163, 170–171, 184, 202, 211, 238, 242

F

falls	105, 166
fatal accident rate	231
fatigue	45–46, 118, 166
of metals	168
fencing (of machinery)	224
Feyzin	55–56
filters	13–15, 30, 54, 61, 109, 177, 209
fires	20, 28, 32, 36, 55, 65–66, 72, 100, 109, 115–116, 120, 125, 156–157, 166, 172, 175, 183, 191, 193, 199, 244, 261
fire–fighting	44, 85, 222
flame arresters	99, 184
flameproof equipment	178
flare stacks	45, 91
Flixborough	66–67, 121, 163, 215
foam–over	239
force and pressure confused	54
forgery	222
forgetfulness	11–46
forgotten hazards	53, 62, 64, 72, 109, 119, 123–124, 164, 184, 234–253, 264
Freud	12
friendly design (see user-friendly design)	
full–scale deflection	91
furnaces	29, 90, 170, 244

G

gaskets	180, 252
'glass boxes'	209
good engineering practice	168
gravity	2
grey hairs	4

H

habits	84–85
hangers	176
Harrow	39

hazard and operability studies (Hazop)	20, 61–62, 91, 155, 160, 165, 205, 207–210, 212–213, 232, 261
hazards, hidden (see also forgotten hazards)	113
Hazop (see hazard and operability studies)	
Health and Safety at Work Act	229
Herald of Free Enterprise	117–118
Home Guard	48
hoses	32, 61, 187–189, 193–194, 200, 243, 265
hydrogen fluoride	202
hydrogen sulphide	78
hygienic factors	258–260

I

ICI	1, 231
identification of equipment	180, 187–189, 235, 248, 250
ignition	114, 116, 171, 199, 261
ignorance	53–67, 103, 164, 175–180, 183, 265
indicator lights	17–18
information	267
information display (see also labels)	126, 156
information overload	80–81
information retrieval	124
inherently safer design	52, 120–121, 151–152, 162–163, 165, 206, 265–266
initiative	5, 98, 106
insects	49
inspection	56, 125, 168, 173, 178–179, 261
instability (of processes) (see also runaway reactions)	101–102
instructions	2–7, 15, 20, 48–76, 107, 113, 119, 192–193, 206–208, 230
contradictory	65, 73

lost 53, 56
ways of improving 67–77
instrument failure 150
insularity 114
intention changes 182–183
interlocks (see also trips) 15, 18–20, 22, 41, 204–206, 214–216, 266
inventories, need to reduce 121

J

joints 34, 55, 114, 178–180, 183, 193
judgement, premature 88–89

K

keys (see buttons)
kinetic handling 75
King's Cross 114, 116
knowledge, failure to use 10, 54–56
knowledge–based tasks 5–6, 72–74, 144, 208

L

labels 20, 26, 28, 30, 53, 60, 80, 248, 250, 252, 263, 265
missing 21–22, 116
ladders 30, 79, 105, 249
lapses of attention vi, 3–4, 8–9, 11–46, 201, 227
law 225–226, 264
layout 22–23, 75, 79, 80, 115, 262
leak testing 187–191
leaks 31–32, 34, 58, 62, 109, 114–115, 121, 159–160, 176, 178, 182, 193–194, 199, 201, 211, 215, 261
onto water 81
lecture demonstrations 45
legal responsibility 221–226
libraries 222
lifting gear 43, 92–93, 179
light bulbs 150
liquefied flammable gas 100, 190

liquefied petroleum gas
(see also propane) 180
liquid transfers 192
lithium hydride 61–62
Lithuania 216
lost–time accident rate 112–113, 230, 260
lubrication 150, 201

M

machines and people compared 83–84
maintenance 4, 17, 27–29, 99, 175–185
avoiding the need 185
of instructions 71
isolation for 17
preparation for 16, 27–29, 32, 72, 81–82, 103–104, 115, 148, 180–183, 187–192, 215, 250
quality of 63, 149–150, 183–185
management 2, 5
education 113, 125
errors 7, 112–125
managerial responsibility 227–231
managers
responsibilities of 112–125
senior, actions possible by 120
manpower, reductions in 60, 67, 122–123, 178
manual handling 75
materials of construction 150, 156, 161, 169–170, 172, 201
effect of temperature on 54, 58
measurement errors, probability of 146
measurement of safety 125
Meccano 3
mechanical handling 198
medical errors 31, 34–35, 214
Melbourne 57
memory (see forgotten accidents)
mercenaries 110
methanol 193
methods of working 17
mind–sets 38, 88–94, 163, 180, 192

mirror images   27
mismatches   v, 5, 78–94, 119
missiles   205
mistakes   v, 5–6, 48–76
modifications   91–92, 160, 172,
                         206, 217, 264
    people not told   58–59, 214
    unauthorized   214–216
monitoring   6
Moorgate   39
morale   109–110, 258–260
motivational appeals   107
motivators   258–260
music   269–270
myths   261–267

N
naval warfare   73, 83
near misses   126
negligence   266
nitrogen   171, 187–189, 237
    blanketing   164–165, 204
non–compliance (see violations)
normal accidents   266
notices   20
nuclear power stations   50–52, 101–102,
           151, 185, 201–202, 216, 266
numbers, errors in   30–31
nuts and bolts   179, 185

O
old equipment and software   217
open containers   236
opening equipment under pressure 13–16
opening full equipment   209
options for the future   164
organizational failures   101, 112
overfilling   189–190, 193, 242
overheating (see also
    runaway reactions)   55, 101–102, 156,
                         179, 196, 245

overpressuring   13–16, 54, 60, 64,
                         91, 100, 156, 159
    probability of   132–134
oxygen   163

P
pacemakers   214
Paddington   40
paraffin   163
parchments   93
patience   94
performance shaping factors   136
permits–to–work   16–17, 81–82,
           103–104, 125, 148, 171,
        179, 180–183, 187–192, 198
personality, effect on accidents of   86
petrol   63, 80, 200
    green   63
phonetic alphabet   198
phosgene   169
pictorial symbols   196
pigs   54
pilot error   41
pipe supports   168–169
pipebridges   33
Piper Alpha   120
pipes   53–54, 60, 66, 150, 156,
              168, 217, 240, 246
    plastic   58
pitfalls   147–149
poisoning   45, 196
policy   120
    statements of   100
polyethylene   31–32
powder   61
preparation for maintenance   250
pressure, excessive
    (see overpressuring)
pressure vessels (see also reactors) 13–16
probability data (see errors,
    probability of)
problems, disappearance of   255
propane   55

prosecution 225
protective clothing 58, 104, 181–182, 238
protective equipment (see also alarms, automation, interlocks, trips) 4, 38, 121–122, 162–163, 165, 189–190, 225
   isolation of 155
protective systems 53
protoplasm 255
psychogenic illness 87
pumps 20–21, 53, 62, 82, 90, 182–183, 201
   automatic start 141–142
pushbuttons (see buttons)

Q
Quintinshill 35–38

R
railways 32, 35–40, 45, 80, 82–83, 85–87, 104–105, 147, 184–185, 198– 201, 228, 253
   station fire 116
Rasmussen Report 136, 138–139
reactions 30, 32
reactors (see also nuclear power stations) 18–20, 25, 27–30, 63–65, 72, 100, 163, 179, 192–193, 207, 211, 212
record sheets 62, 91
recovery from error 1, 27, 146, 149
refrigeration 91
reliability, high 149
relief valves 20–21, 51, 55, 57, 90–91, 104–105, 156, 159, 182, 184, 190–191, 193, 252–253, 266
repeated accidents (see forgotten hazards)
research 10
respirators 30
responsibility
   legal 221–226
   managerial 227–231
   personal 221
risks, tolerability of 151–152
road accidents 43
road tankers 61
roofs, falls through 105
root causes 122–123
rule–based tasks 5–6, 72–74, 144, 208
rules, violations of 97–110
rules (see also instructions) 208, 264
   relaxation of 66
runaway reactions 30, 32, 53, 63–65, 207–208, 211
rupture discs 251

S
safety advisers 264
safety management systems 122
Salem 94
sampling 238
scaffolding 54
scapegoats 73–74, 227, 230, 263
scenes of accidents, visits to 105
scrap 241
'second chance' design 160
self–inflicted injuries 87, 258
Sellafield 159–160
seniority 116
service lines 116
settling 160
shipping 117–118
short cuts 36, 104, 181–183, 189
signalmen's errors 35–38, 83, 85–86, 198, 200–201
signals passed at danger (SPADs) 38–40, 80, 82–83, 86, 147
similar names 198–199
simplicity 22, 121, 163
'skill of the craft' 161, 178
skill–based tasks 5–6, 72–74, 144
slips vi, 3–4, 8–9, 11–46, 90, 118–119, 205, 213, 227, 266
smoking 103

software, inscrutable 209
software tests 205–206
space rockets 217
SPADs (see signals passed at danger)
Spain 190
spares 106, 109
spelling errors 7–8
spillages 191, 204, 209, 237, 253
static electricity 61
steam 13, 60, 91, 180, 201, 240
steam drums 22
stirring (see agitation)
stores, unofficial 106
stress 12, 26, 43, 130, 142–143, 146, 166
stress concentration 161
supervision 37, 118
supervisors, overloading of 81
system failures 101

T

tailors 9
tankers 63, 189–192, 194, 200, 237, 243
tanks 56, 70, 80, 91, 100, 160, 171–172, 184
overfilling, probability of 143
task overload 81
task underload 83–84
tasks, difficult 98
team working 60
telephone numbers 43
temperature, effect on materials of 54, 58
construction
TESEO 136–137
tests 4, 21–22, 62, 72, 100–102, 104–106, 119, 237–238, 242
half–done 105–106
software 205–206
theft, avoiding 222
THERP (Technique for Human
Error Rate Prediction) 136, 140
Three Mile Island 50–52

time 263
effect on errors 144
tolerance (of misuse) 121, 173
'track circuits' 37–38
train driver errors 38–40, 80, 82–83, 86, 104, 228
probability of 147
training 2–8, 11–12, 48–76, 84, 113, 124, 168, 173, 178–179, 181, 189, 193, 215
unavoidable dependence on 143
transfers of liquids 156, 159–160
trapped liquids 109, 175
trapped pressure 176, 183, 194, 243
traps 26, 41
trips (see also interlocks) 21–22, 32, 50, 100–101, 125, 131, 155–156, 190, 204–206, 263, 266
turbines 50
tyres 156

U

unconscious mind 11–12
unforeseen hazards 234–253
unforeseen problems (see also
modifications) 211–212
university degree courses iii
user–friendly design 3, 52, 120–121, 206, 212, 248, 249, 251–252, 266

V

vacuum 62
valves (see also
drain valves) 13–24, 22–23, 27–29, 65, 79, 109, 131–132, 151, 175, 177, 180, 182, 190, 199, 212, 214–215
vehicles (see also cars) 84–86, 156, 198, 245, 262
ventilation 172, 178

vents   20–22, 63, 79, 90, 184, 193–194, 236, 239, 246

verbs, irregular   98, 265

vessel failure   55, 159, 192–193, 212, 246

vessels, sucked–in   63, 184, 239

violations   v, 5–6, 8, 97–110, 118–119, 214–215, 227–229

    by managers   99–102

    by operators   103–106

    can prevent accidents   106

    prevention of   97–98, 107–109

    of rules   97–110

viruses   216

visibility   78–79

W

waste heat boilers   22, 74

waste material   202

watchdogs   204–205

watchmen   83

water   50–54, 172, 178, 237, 246

    electrolysis of   104

water hammer   60

weld failure   201

witches   94

wrong method, right answer   267

Y

Yarra river bridge   57

Yugoslavia   227–228